Home Recording Studio

Build It Like the Pros

Second Edition

Rod Gervais

Australia, Brazil, Japan, Korea, Mexico, Singapore, Spain, United Kingdom, United States

**Home Recording Studio: Build It Like the Pros
Second Edition**
Rod Gervais

**Publisher and General Manager,
Course Technology PTR:**
Stacy L. Hiquet

Associate Director of Marketing:
Sarah Panella

Manager of Editorial Services:
Heather Talbot

Marketing Manager: Mark Hughes

Executive Editor: Mark Garvey

Project and Copy Editor: Marta Justak

Technical Reviewer: Gino Robair

Interior Layout: Jill Flores

Cover Designer: Luke Fletcher

Indexer: Sharon Shock

Proofreader: Michael Beady

© 2011 Rod Gervais.

ALL RIGHTS RESERVED. No part of this work covered by the copyright herein may be reproduced, transmitted, stored, or used in any form or by any means graphic, electronic, or mechanical, including but not limited to photocopying, recording, scanning, digitizing, taping, Web distribution, information networks, or information storage and retrieval systems, except as permitted under Section 107 or 108 of the 1976 United States Copyright Act, without the prior written permission of the publisher.

For product information and technology assistance, contact us at
Cengage Learning Customer & Sales Support, 1-800-354-9706.

For permission to use material from this text or product,
submit all requests online at **cengage.com/permissions**.
Further permissions questions can be e-mailed to
permissionrequest@cengage.com.

All trademarks are the property of their respective owners.

All images © Rod Gervais unless otherwise noted.

Library of Congress Control Number: 2010932780

ISBN-13: 978-1-4354-5717-1

ISBN-10: 1-4354-5717-X

Course Technology, a part of Cengage Learning
20 Channel Center Street
Boston, MA 02210
USA

Cengage Learning is a leading provider of customized learning solutions with office locations around the globe, including Singapore, the United Kingdom, Australia, Mexico, Brazil, and Japan. Locate your local office at: **international.cengage.com/region**.

Cengage Learning products are represented in Canada by Nelson Education, Ltd. For your lifelong learning solutions, visit **courseptr.com**.

Visit our corporate Web site at **cengage.com**.

Printed in the United States of America
2 3 4 5 6 7 12

To all five of my children,

I want you to know that I am proud at the sense of drive and ethics you have brought with you into adulthood, and to see you raising your children with those same values (well, those three of you who have children anyway).

I would ask you all to remember that your dreams should not be set aside and forgotten, gathering dust in some chest in your attic, but are meant to be embraced, fed with fuel for their fires, so that you attain (in your lives) any and everything you could ever desire.

The only possible obstacle in achieving this is you. Don't ever let that be the case.

Acknowledgments

I wish to acknowledge the following people for their assistance with this book:

Executive Editor Mark Garvey, for putting together everything it took to make this happen.

My editor, Marta Justak, who again put up with me while we put this together. Marta, you truly are an angel.

Technical editor, Gino Robair, your efforts have made this edition of the book much more than it would have been otherwise.

To the compositor of the book, Jill Flores, I know you jumped through a lot of hoops making it from A to B in this edition; I appreciate this more than you could ever know.

Doug Plumb of Acousti Soft Inc., thanks for your assistance with Chapter 8.

Jeff D. Szymanski, PE, special thanks for your assistance in general, as well as providing review for technical accuracy.

Brian Ravnaas, thanks for your assistance and contributions to Chapter 3.

To my brother Paul, the photographer, for providing my photo. You did a great job of making me look much better than I do in real life.

A special thanks to my readers, the people who made the first edition such a success that we actually had a reason for releasing the second edition. It amazes me the support you have provided to me—both from your warm reception of the book to your recommendation to people that they purchase it. Without you, this edition really would have never happened.

And finally, last but certainly not least, again, to my brother Marc. You have been a source of strength to me ever since we were kids—a shining light I could follow. I notice even this late in life that nothing has changed.

About the Author

Rod Gervais is a multi-disciplined engineer located in northeast Connecticut. He is a member of the Acoustical Society of America, the Audio Engineering Society, and the National Fire Protection Association. His background in construction is wide ranging, from national museums to recording and movie studios. It was his involvement in recording studio construction that first fired his need for knowledge in the field of acoustics, a fire that still burns today. His quest for knowledge is a never-ending part of his life.

Rod is also an accomplished musician, with over 40 years of experience in percussion, as well as playing guitar, bass guitar, and keyboards. His musical experiences run the gamut from classical to jazz, although the musical love of his life is the blues.

It is this combination of the analytical engineer coupled with the artistic musician that has led him to write this book—one more step on his journey through life.

Contents

Introduction xii

Chapter 1 Getting Started 1
    A Variety of Equipment Options 1
    The Room's Design 2
        Power Station—the "A" Room 3
        Hit Productions—the "H" Room 10
        Dark Pine Studios 12
    Sound Isolation 15
    What Sets This Book Apart from All the Rest 16
        Easy Explanations 16
        Thirty-Five Years of Construction Expertise 17
        The Devil Is in the Details 18
    Ensuring Quality Control Protection 19
        Doing the Job Yourself 19
        Hiring the Job Out 20

Chapter 2 Modes, Nodes, and Other Terms of Confusion 22
    Sound 22
        Amplitude 23
        Frequency 23
        Wavelength 23
    Room Modes 24
        Modal Waves 25
        Non-Modal Waves 28
    Other Reflective Problems 29
        Flutter Echo 29
        Comb Filtering 30
        Early Reflections and Stereo Imaging 31
    Mode Analysis 32
    Room Sizes 33
        Large Rooms 33
        Medium Rooms 33
        Small Rooms 33

Types of Room Modes 34
Axial Modes 34
Tangential Modes 35
Oblique Modes 35
Room Ratios 36
Mode Calculators 37

Chapter 3 Isolation Techniques—Understanding the Concepts 41
Mass, Mass, and More Mass 43
Airtight Construction 46
Eliminate Transmissions Through the Building Structure 46
What to Avoid 47

Chapter 4 Floor, Wall, and Ceiling Construction Details 48
Floor Construction 48
Simple Concrete Slabs 48
Isolated Concrete Slabs 49
Floating Concrete Slabs 52
Description of Test Specimen 56
Test Methods 56
Test Configuration 57
Test Results 58
Floating Wood Decks 60
Sand-Filled Wooden Decks 62
Wall Construction 62
Existing Walls 64
Wood Walls 65
Steel Framing 71
Masonry Construction 72
Ceiling Construction 73
Working with Existing Ceilings/Floors 73
Resilient Channel Ceilings 78
Suspended Ceilings 82
Semi-Independent Frame Ceilings 82
Independently Framed Ceilings 85

Contents

    Additional Isolation Products 87
    Damping Systems 88
    Optional Systems 91

Chapter 5 **Window and Door Construction** 93
    Glass 93
      Float Glass 94
      Heat-Strengthened or Tempered Glass 94
      Laminate Glass 95
      Plexiglass 96
    Window Frame Construction and Isolating Techniques 96
      Window Frames and Trims 96
      Glass Thickness 100
    Manufactured Window Units 103
    Constructing Doors 103
      Door Frame Construction 104
      Windows in Doors 108
      Door Hardware 108
      Adding Insulating Panels to Door Assemblies 109
      The Finished Product 111
    Manufactured Doors 112

Chapter 6 **Electrical Considerations** 114
    Line Voltage 114
    Low Voltage 117
    Electrical Noise 118
      Ground Loops 118
      Ground Loop Solutions 121
    Isolated Ground Receptacles and Star Grounding 122
    Lighting 124
      What Is Radio Frequency Interference (RFI)? 125
      The How's and Why's of Lighting Noise 126
    Diagnosing and Troubleshooting Problems 126
    Some Common Problems 127
    Fixing the Problem 127

Chapter 7 HVAC Design Concepts 130

- Getting Started 130
- Room Design Criteria 131
  - Btu Output 131
  - Why Should I Care About Humidity? 136
  - Understanding the System as a Whole 139
- System Options 140
  - Split/Packaged Direct-Expansion (DX) Air Conditioners 140
  - Through-the-Wall Systems 142
  - Ductless Mini-Split Systems 144
  - Portable Air Conditioners 144
  - Evaporative Coolers 145
  - Exchange Chambers 146
- Combination Cooling/Heating Systems 148
  - Split Systems 148
  - Through-the-Wall Systems 150
  - Exchange Chambers 151
- Separate Systems 153
- System Design 154
- Noise-Level Design Guides 158

Chapter 8 Room Testing 161

- Getting Down the Basics 161
- Sound 162
- The Software 162
- Room Anomalies 163
  - Early Reflections 164
  - Resonant Sounds 165
- Tools of the Trade 166
  - Impulse Response 167
  - Frequency Response Curve 168
  - Gating 169
  - Waterfalls 170
  - Psychological Response 175
- Proper Use of the Software 176
  - The Nature of Scientific Measurements and Experiments 176
  - Hardware Connections 178

Contents

      Data Gathering 179
      About Signal-to-Noise Ratios 182
      Operation of Analyzers 183
  Parametric EQ 190

Chapter 9 Room Treatments 194

  Low-Frequency Control 194
  Pressure Devices 196
  Velocity Devices 196
  Hybrid Devices 197
  Early Reflection Control 197
  Absorption Coefficient 199
  DIY Treatments 201
      Fiberglass Panels 202
      Low-Bass Panel Traps 208
      Helmholtz Traps 210
      Mid/High-Bass Absorbers 214
      Diffusors 215
  Manufactured Treatments 226
      Auralex 228
      RealTraps 231
      GIK Acoustics 233
      Ready Acoustics 236
  So Where Do We Go from Here? 237

Chapter 10 Putting It All Together 239

  Studio Design and Detailing 240
      On to Options 245
      HVAC Systems 251
      Developing the Details 254
  A Few Different Case Scenarios for Your Perusal 269
      An Inexpensive but Effective Decoupling Floor System 269
      Beefing Up Garage (or Other Exterior) Wall Assemblies 271
      Attics and Other Room Spaces Inside of Your Home 274
      Studio Treatments 290
  The Finished Product 299

Chapter 11 Myths and Legends 300
Some Popular Myths and Legends 301
Fiberglass 301
Egg Crates and Other Great Acoustic Treatments. 303
Please Buy Our "Soundproofing" Materials 312
Close Is Good Enough 313
The Final Word on Myths 317

Chapter 12 Codes, Permits, and Special Needs 318
Building Officials 318
Structural Analysis of the Proposed Work 319
Problems Relating to the Bearing Capacity of Earth 319
Unstable Conditions Caused by the Addition of Bearing Walls 323

Glossary 328

Appendix 340

Index 342

Introduction

Shortly after the release of the first edition of this book, I set up an email site for those who might want to comment on the book. One of the things I asked my readers was for them to send along a wish list of things they might want to see added in a future edition.

Although I am not going to include everything that was requested (one example being adding construction terminology to the Glossary), I have made some major changes to this edition.

I've made minor revisions to Chapter 1, "Getting Started."

In Chapter 4, "Floor, Wall, and Ceiling Construction Details," I have added some wall isolation details, a method of isolation for floors that I developed after the original book, details for typical garage installation, and a number of details you asked for. I have also included some information on concrete and masonry structures for our friends overseas, who don't work with wood as much as we do here in the States.

In Chapter 6, "Electrical Considerations," I've added quite a bit of material, including low-voltage wiring, LED lighting, and various wiring methods.

Chapter 7, "HVAC Design Concepts," includes additional details covering duct linings and methods of isolation chamber design.

Chapter 8, "Room Testing," has been rewritten in its entirety and now features the RPlusD software package from Acousti Soft, Inc. This is the same company that produced the EFT Software reviewed in the first edition. I would also point out that Doug Plumb assisted me in writing this chapter and did the final review to make certain it represented his product accurately.

In Chapter 9, "Room Treatments," I added various DIY treatments.

In Chapter 10, "Putting It All Together," you'll find additional details to help you find your way.

All the remaining chapters will be unchanged.

Again, I tried to hold true to the original concept of the book, which included no math-heavy problems to deal with and a hands-on approach to the people who only want to "get it done" so they can make their music.

Companion Web Site Downloads

You may download the companion Web site files from www.courseptr.com/downloads.

CHAPTER 1

Getting Started

So…you want to build a home recording studio? Well, join the club.

Due to the recent advances in technology in the field of electronics, building a home recording studio is now more of a reality than it was 15 years ago. For example, 15 years ago, setting up a home studio to record 24 tracks at once would have been unthinkable for most people. The cost of the gear alone would have made it almost impossible to do, unless someone had some serious money to burn for a hobby. Today, though, the *unthinkable* has become the *affordable*.

A Variety of Equipment Options

Just look at current gear costs for a moment. For an investment of about 6,000 U.S. dollars, you can have a 24-track board designed by Malcolm Toft of Toft Audio Designs, which has the same technology used on the Trident Boards he designed back in the '70s and '80s. Better yet, it's constructed directly under Malcolm's watchful eye. That price is amazing when you think about the fact that today you can spend upward of $27,000 for a used 32-track Trident Console. Heck, even the 80B input modules cost $950 a pop used—*if you could find them.*

Plus, you can have top-of-the-line tube gear produced by companies like Manley in Chino, California, or Sebatron in Australia. EveAnna Manley and Sebatron themselves are personally leading the way to innovative designs, since they directly oversee the construction of the gear they're selling you.

Then you have thousands of options for affordable solid state gear, including effects boxes, compressors, limiters, pre-amps, amplifiers, and so on. Companies like Joemeek, Tascam, PreSonus, M-Audio, and more are all giving you a wide variety of options for your recording needs.

All of this creates an affordable cost for starting your own home recording studio.

The Room's Design

Most of you reading this book probably have a fair idea of how to use this gear—after all, you're musicians. However, what you lack is the knowledge it takes to build a room in which you can use this gear effectively.

Don't get me wrong—you can set up in any old room of your house (or apartment) and play around with your gear. Maybe you can even make some halfway-decent-sounding demo CDs if you run back and forth 10 or 12 times to your car so that you get the levels corrected. But without a decent room, you'll never be able to sit down and really make anything you'll be truly proud of and excited to have other people hear.

The gear is only part of the equation when it comes to recording and mixing; the other parts are the rooms you do it all in. In this chapter, you will see a world-class recording studio I had the pleasure of "gently guiding" from a hole in the ground to final finishes. It was designed by acoustical genius Tony Bongiovi. It was the first recording studio I ever worked on, and it was the beginning of my love and appreciation for everything special that it takes to make a recording studio.

Some Background on Tony Bongiovi

Tony Bongiovi is the president and co-founder of Bongiovi Entertainment and has built a "solid gold" reputation as an engineer and record producer, amassing more than 50 gold and platinum albums since beginning his historical career at the famous Motown label of the late 1960s. His production discography alone includes such artists as Jimi Hendrix, Bon Jovi, Talking Heads, The Ramones, Ozzy Osbourne, Aerosmith, Gloria Gaynor, and The Scorpions. He co-produced the biggest selling pop instrumental single of all time, "Theme from Star Wars" by MECO.

Tony conceived, created, and operated the Power Station Studios, the New York–based mega-studio (now Avatar Studios) whose rooms remain one of the gold standards in the recording industry. Power Station's client list reads like a "who's who" of the music industry.

I am also going to show you a couple of studios I designed, so you can get a feel for options that exist.

Designing and constructing a recording studio is not like any other construction you might do. If you begin with a blank piece of paper and an empty piece of land, designing and constructing a home is easy. You decide the type of home you want, the total square footage you can afford, and then you make it all fit.

If you have to shift a wall 6" to fit some furniture in—no problem. If a bedroom ends up being 10x10—no problem. You only have to worry about the house being functional, but never have to concern yourself with how it will sound. Yes, I said "sound," because sound is a function of room size. You'll learn more about this in the next chapter.

With a recording studio, not only do you need to concern yourself with its function, but you also have to concern yourself with its sound. That same 10x10 bedroom that isn't a problem for sleeping or reading is a problem for recording. And if it had 10' ceilings, it would be even worse.

Next, you have to be concerned about what function you will be performing in the room—because function drives design. For example, if you are going to have a dedicated control room, then you want the room to be perfectly symmetrical. Whereas, for a dedicated tracking room, asymmetrical would be ideal. With a combination tracking/control room, settle for symmetry; otherwise, the control room part of the equation won't work.

Take a look at the figures that follow. These very different rooms in Studio "A" have different needs and therefore different finishes, shapes, and sizes.

Power Station—the "A" Room

Power Station is one of the premier names in the recording industry. In the main tracking room of the facility (see Figure 1.1), you can record anything you choose, whether it's a full band or just a vocal. At ground level, it (the sound in the room) is fairly dead—mics at that level capture just the instruments. The higher you place the mic in the room, the more lively it gets. By the time you reach the "coffin" you see at the top of Figure 1.1, it sounds as if you are recording in a nightclub. A combination of close micing and room mics for ambiance make for great recording from this room.

"The Coffin"

The "coffin" I refer to is actually the coffin-shaped compression ring that the main structural members bear on to form the ceiling frame. You can see it in the very top of the room in Figure 1.1. We refer to it as "the coffin" because it is coffin shaped. Due to its size, shape, and location, it is acoustically very active.

All of these needs (and effects) were considered by Tony as he worked his way through the design of this studio in his mind, but they were never perfectly executed until this particular studio was constructed, which is why *this* studio did not require "tuning" upon completion.

Figure 1.1
Main room.

Notice how the board spacing tightens up (the space between boards gets smaller) as you climb higher into the room. This creates a greater ratio of board surface to airspace between the boards (and whatever is behind the boards). The more square feet of exposed board (per square foot of wall surface) you have, the more the mid and higher frequencies are reflected.

Also, as the room rises, notice how it closes down until it finally reaches the smaller volume of the coffin. This design effectively creates a gradually greater reverberant field as you rise into the room. That's why mics placed in the coffin capture that exciting, vibrant feeling, as if you recorded live in some nightclub.

I can't tell you any more about this than I have. The items discussed here are all visible on the surface, and anyone who is an expert in acoustics would be able to see and understand them. But there is much more behind the boards that comprise this work as well—from the designed construction of the walls to their geometry in relation to the geometry of the finished surface. All of this works together hand-in-hand to create a room like this.

Room Geometry

The geometry at the face of the finishes in this room is not the same as the geometry at the face of drywall behind them. The geometry of the room itself (pre-finishes) is important from the perspective of controlling modal activity. You will examine this more in Chapter 2, "Modes, Nodes, and Other Terms of Confusion." The finishes (on the other hand) are the manner in which you fine-tune the room to cancel the remaining acoustic anomalies. The two do not necessarily have to be anywhere near the same shape in order to end up with a room that is acoustically pleasant.

In the percussion/rhythm room (shown in Figure 1.2), you see just the opposite. The board spacing is wide and very open between the boards. This lowers the ratio of mid/high frequencies reflected back into the space. Cymbals will have a faster decay in this room, while the low-frequency thump of the bass drum will be accentuated. Bass guitar recorded in this room is clear and concise once again, with all of this carefully controlled by what you see, as well as what you don't see. By the way, modal issues aren't a problem in this space because of the way it opens into the main room. There is an isolation assembly for this room, but it uses glass partitions that allow just enough isolation not to cause problems in the main room, while not creating modal issues in this room. As you will see, that's quite a juggling act.

Figure 1.2 Percussion/rhythm room.

The string room (see Figure 1.3) is designed to record a grand piano, full string section, or a single violin. This design begins with slightly tighter board spacing at floor level, becoming increasingly tighter as you go higher in the room. It's a relatively small space with a tall vaulted ceiling. Oh, one thing to remember as you work your way through your studio, don't be afraid to get creative. In the original Studio "A," there was no chance for what happened here, so it did not exist in the original. But in this studio, the wall you see on the left side of the picture was adjacent to the building's exterior, with just enough room for an air-lock entrance. It's a great place for vans to back up and enter directly into the room. But there were no doors in the original, so we put some in and made them almost invisible. If you look carefully, you can see the edges of a pair of 3' doors by the corner of the two walls, with the surface finish applied over them.

Figure 1.3
String room.

Changes to the Power Station Design

We designed these doors in the manner we did because of the psychological effect that making any major changes to this studio could have brought into play.

There are people who have worked in the New York rooms that were totally convinced that any changes to these rooms would have rendered them inferior to the original rooms. It would not matter (to them) even if you could provide hard data through testing that proved the rooms were acoustically superior, because in their minds they would be inferior.

So that you can better understand this I offer the following:

Tony had a control room in New York that had a brick wall immediately adjacent to one wall of the room, and the wall opposite of this wall contained the entrance to the control room.

Tony was always very careful to design his control rooms so they were acoustically symmetrical. This room was no different. However, there was a whole group of engineers who would not mix in that room because of the physical differences between the left and right walls. And, no matter how much Tony tried to explain that the room was acoustically balanced, they insisted that they could "hear" that it wasn't.

Tony finally installed a door assembly with all associated hardware in that wall, at which point in time the "problem" was solved.

Acoustically there was no problem, but you could never convince those people of that. If at all possible, design your control rooms so that you achieve both acoustical and physical symmetry.

Figure 1.4 gives you a different view of this room (looking into the main room), with the isolation panels in the closed position. There are three sets of these movable partitions that separate this room from both the main and rhythm rooms and the rhythm room from the main room.

Figure 1.4
A different view of the string room.

In the iso-booth, you can see some very different construction, as shown in Figure 1.5. Acoustic ceilings come into play. It still has wood-finished walls and floors, but in this room, you want the sound to be slightly more dead than in the other rooms. You don't want a very reverberant room, but rather, you want to capture more of the vocals or instruments. Adding acoustically absorbent materials to the ceiling helps a lot with this cause. These rooms are not made for low-frequency instruments like bass guitars, which would cause serious modal issues in a space this size. Note how the walls in this room are not parallel, which helps to avoid issues with flutter echo and comb filtering. You'll learn more about that in Chapter 2.

Figure 1.5 Iso-booth.

Finally, you can see the control room, as shown in Figure 1.6, and note that once again the design is very different from any of the other spaces in the studio. This room is symmetrical, whereas the tracking rooms are not. Symmetry in a control room is critical to proper stereo imaging. As you saw in the iso-booth, the control room has a soft absorbent ceiling for acoustic control. Note how the ceiling is splayed from a low point over the board to high points at the front and back of the room. Again, this helps to alleviate problems with comb filtering and flutter echo at the engineer's location.

Figure 1.6
Control room.

This room is designed to approach the acoustics found in a typical living room, but without the acoustic anomalies that would exist there normally. If you look at the rear wall of this room (see Figure 1.7), you'll see that this is also laid out in a symmetrical manner, with plenty of soft fabrics (and what's behind them) to help tame early reflections and modal issues within the space. Plenty of space is provided for rack gear and patch bays on this wall. Front to back, left to right—perfect symmetry equates to perfect stereo imaging.

Figure 1.7
Rear wall of control room.

Looking through this world-class pro studio gives you a feel for what is involved in studio design and construction, but unless you have some serious money to invest, you aren't going to go to this extreme in your home studio.

Hit Productions—the "H" Room

Hit Productions is an award-wining Audio Visual (AV) firm. Studio H was their first venture into the world of real-life audio recording. In the control room of this studio (which was constructed in the Philippines), you can see a design completely different from Power Station. Whereas Power Station has a compression ceiling (the ceiling is lower over the board and higher in the back and front of the room), this ceiling is sloped from a low point in the front to a high point in the back of the room. This is also not a wood-finished room. The owner wanted a more modern look with some fabric finishes. Note the ceiling cloud hung below the actual ceiling (see Figure 1.8).

Figure 1.8
Studio "H" control room.

In the back of the control room (Figure 1.9), you have the entrance door on the right and a storage room on the left (which also houses the studio computers). In between those doors and along the ceiling above them is the front finish for the room treatments handling the low frequencies. Given enough area in the footprint and careful planning, room treatments do not need to be an afterthought.

The treatment space behind those finishes is about 6' in depth. Chapter 9, "Room Treatments," will include details outlining how you can effectively use a space like that in your construction.

Figure 1.9
Studio "H" control room rear wall.

The owner (due to space considerations) opted for a string room (which can double for an iso-booth) in lieu of a typical iso-booth (see Figure 1.10). Although relatively small in area, it has fairly high ceilings that are installed to create a rather dramatic barrel roll effect—you can see this clearly on the upper-right side of the room in Figure 1.10.

The finish in this room is all wood, which helped achieve very bright high- and mid-frequency responses, while maintaining good low-frequency control. That baby grand piano you see in there sounds sweet—a string section that recorded in there fell in love with the room.

Figure 1.10
Studio "H" string room.

Notice that the tracking room in this studio (Figure 1.11) is much smaller in its footprint than the one at Power Station, because the building housing this studio had nowhere near the necessary clearance to achieve the ceiling heights you see in Power Station's tracking room. However, we were able to work to a high point of 17' on one side of the room, which gave us everything we needed to make a room with first-rate sound. Again, tracking rooms are best when they are asymmetrical, which is what you see here.

Figure 1.11
Studio "H" tracking room.

I have a couple of comments regarding this studio. The outside of the studio itself is exposed to common spaces, corridors to be exact. This studio is housed on the third (top) floor of a commercial building in the heart of Manila (Makati City to be exact), and to make things more difficult, it has a tin roof, which is a challenge from an isolation point of view. We will examine (in Chapter 10, "Putting It All Together") how you can deal with conditions like this.

All of the air-handler units and duct work for the air supply were installed between the studio ceiling and the tin roof, but the return air was only installed outside of the rooms (above the hall ceiling), which involved another isolation challenge (see Chapter 7, "HVAC Design Concepts," for more complete details).

Dark Pine Studios

Dark Pine Studios is a smaller production studio that is fast making a name for itself in the recording industry. In the control room of this studio (which was constructed in North Carolina), you see another control room design that differs from Power Station (see Figure 1.12). Again, where Power Station has a compression ceiling, this ceiling is sloped from a low point in the front of the room to a high point in the back of the room. Similar in some respects to Power Station, it is a wood-finished room because the owner wanted the warmer feeling that wood gives.

Figure 1.12
Dark Pine Studios control room.

In Figure 1.13, you can see the back wall of the control room. Note that it is similar to Power Station in that it has gear racks built into the wall.

Behind this space you can find a lounge and a computer room. Because the owner was not planning on using these spaces while recording, we got creative and designed the back wall of the control room with little to no isolation. This effectively created a room that was much deeper than it seems at first glance, which allowed us to use this space as a deep bass trap.

You can also see (at the top of the wall) a soffit containing the return air outlet, as well as a couple of supply registers. In between these spaces (within the soffit), we developed deep bass traps. Both of the vertical corners have bass traps as well. You can see one of those on the left side of the picture. We will discuss methods of constructing these types of room treatments in Chapter 9.

Also note the lack of a ceiling cloud in this room—all that was needed acoustically in this case were simple 12x12 acoustic ceiling tiles.

You can see that the tracking room in this studio (Figure 1.14) is also much smaller in its footprint than Power Station. The building housing this studio had nowhere near the clearance necessary to achieve the ceiling heights you see in Power Station's tracking room. However, we were able to work to a high point of 12'6" on one side of the room, which gave us everything we needed to create a room with first-rate sound. The asymmetrical design of this room is completely different than either Power Station's "A" room or Hit Production's "H" room, yet they all have a warmth and clarity that make them a pleasure to record in. Note the small iso-booth in the back of this room.

Figure 1.13
Dark Pine Studios control room rear wall.

Figure 1.14
Dark Pine Studios tracking room.

As I mentioned earlier, you probably won't be investing as much money or labor in your home studio as you see in the pro studios above. The idea here is to show you that there are a lot of different approaches you can take with design while accomplishing the same thing—a room that has a great sound.

Now, let's take a quick look at some of the challenges facing you in your quest.

Sound Isolation

You can't forget the issue of sound isolation, which is possibly one of the most important aspects to consider when designing a home studio, and even high-end residential construction is generally not good enough to satisfy those needs.

In any house I design, we generally put sound-batt insulation in every sleeping area and bathroom. We also use resilient channel with sound-batt insulation for the master bedroom. This will generally make it so that you can't listen to conversations taking place in those rooms, which represents a decent level of sound isolation for a home.

What Is a Sound Batt?

Sound-batt insulation is a lightweight, fluffy fiberglass insulation placed inside wall cavities to help stop the passage of conversations from one room to the next. This material is less dense than the standard fiberglass insulation typically used to keep a building's interior environment from being affected by outside conditions.

But that doesn't even *begin* to produce the isolation required for a recording studio.

It's one thing to ask a teenager to turn down his stereo because it's too loud. It's another thing altogether to have to turn down a band that you're recording because the baby can't sleep. Or to have someone drop a dish on the floor above you and have the resultant sound destroy an otherwise perfect take in a recording session.

No—standard isolation techniques simply will not suffice in a studio environment.

If you live in a condo, an apartment building, or some other environment where your ability to modify the existing structure is severely limited, you're probably thinking that buying this book was a waste of money because you won't be able to do what you need to create your studio. Actually, this may or may not be true. You may not be able to soundproof your room, but there are plenty of treatments you can use to improve its sound that are nondestructive in nature. You can either build or purchase room treatments and install them in your space to deal with modal and non-modal issues (more about them in the next chapter). Then, perhaps, you can develop a schedule with your neighbors, such as times when they won't be around, where you can work with your band to record. Maybe you can record with headphones on and everyone DI (direct in), including electronic drums (*ouch*—as a drummer not my first choice, but it would beat the heck out of not playing at all) in order to record. Then you could mix down afterward, at a lower volume and at times when no one else is going to be bothered.

If this is the case, then you probably want to go directly to the second chapter, from there to the eighth chapter, and finish with Chapters 9 and 10. Those are chapters you need for treating the room itself—the rest of the book pertains to sound isolation construction techniques along with electrical and HVAC details.

If not, then keep reading.

What Sets This Book Apart from All the Rest

This book is going to show you how to get it done. How to determine the best use of that space you have, how to isolate yourself from the remaining space in your home (and the outside world at the same time), and how to treat it so that not only does your music sound right, but the room does as well.

Actually, there are other books on the market that you can buy that will show you all of this, but there is a difference between this book and those books. Acoustics (whether it be for sound isolation, room modes, or room treatments) is all about math. It is not necessarily intuitive, and that (for a lot of people) is a problem.

Easy Explanations

Not everyone gets along with higher math—for some people, it's like speaking Latin. When they see equations on a page, they start having flashbacks to high school trig classes—with teachers they didn't especially like—and frustrations they don't want to revisit. Their sight begins to get blurry, their heart starts beating faster, they get faint, and they long for a drug to calm them down.

Well, don't worry, because with this book you don't have to deal with math at all. You'll get straightforward, easy explanations about how to make it from point "A" to point "B." I'll break down some of the math into its simplest forms for you, and I'll also provide you with the tools (in the form of a series of spreadsheets available from the author on request), to help you get the job done.

Simply email me at rgervais10@hotmail.com, or stop into any of the forums I visit to ask for a copy of my tool kits.

Thirty-Five Years of Construction Expertise

Another thing that will set this book apart from others is that the people who wrote those other books are not necessarily experts in construction itself, although they may be some of the best and brightest in the world of recording studios and acoustics. With this book, I bring you over 35 years of construction experience, so ideas that don't really work won't be in here. Also, I'll give you some construction techniques that aren't found on paper anywhere else, but have proven themselves to be useful for me over the years.

There's no magic with any of this, regardless of what you might have heard about egg crates. There are, however, some formulas for success that (if you follow them very closely) will give you a home recording studio you can be proud of—one that will give back to you what you expect for your investment in time and money.

One thing I do want to make very clear to you is this: If you don't get it— read it again and again—and then once more until you do get it. Do not stumble blindly forward thinking that you can slip a sheet of drywall in a location where this book says you don't want it. Don't think that leaving a sheet of drywall in a location where the book recommends that you "take it out," is *not* going to make much of a difference. If you're going to bake a cake, then leaving out key ingredients is not going to get you there, and adding ingredients willy-nilly isn't going to get it done either. Follow the recipe or don't bother with the baking.

When Things Go Wrong

In 2003 I joined a Web site called www.recording.org (highly recommended, by the way, for those of you interested in advice for recording techniques by some of the pros in the industry), and shortly thereafter I was asked to help moderate their acoustic forum. In the past seven years, you would not believe the number of times that people have come to the site and asked for advice, only to return later and complain that their rooms did not have any real isolation value. They are totally frustrated, have spent a lot of money building their rooms, and they are not much better off than before they started. They just can't understand it.

So I spend a lot of time going over with them exactly what they did in the construction of their rooms, just to find out that along the way they decided that it would not make a big difference if they just left the existing drywall on the exterior wall in place (or ceiling if they built in a basement).

Well, they were wrong. They were told to remove it—but they couldn't see how it could hurt—so they ignored us.

Now they have a finished room—spent thousands of dollars to build a room within a room—and they can only fix the problem by removing finishes to get at that old drywall and remove it. Pulling out that old sheet can mean the difference of 10 to 20dB of isolation, and now it also means thousands of dollars in re-work for the owner.

It's not only what you have in the construction, but also where you put it that makes the difference between an assembly working or not working.

The Devil Is in the Details

I've had people come to me with the same complaint over and over again—for example, they've spent time reviewing their work and looking at pictures, just to find out that what they installed as resilient channel wasn't resilient channel at all, but was hat channel. After spending thousands of dollars for materials and labor, the answer is to rip it all down and start over. Had they paid attention in the first place, none of this would be necessary, but they didn't and now it's going to cost them.

One important point—please don't trust your contractors to know the difference between isolating and non-isolating materials, because the vast majority of them don't. In one of the cases with hat sections being substituted for resilient channel, the substitution was done by a contractor hired for the project, whose contract specifically stated RC-2, and who argued with the homeowner afterward that there "really wasn't any difference." Well, I have a surprise for him—there really is.

A hat channel is simply designed as a framing member, generally used to frame suspended ceilings in commercial work. It is a cold, rolled steel member without any isolating slots cut into its legs. RC-2 is specifically designed and tested to create isolation between a structural assembly and the wall surface, which is usually drywall. The isolating slots cut into its legs make all the difference. It's just that simple. That's one of the reason why you will find illustrations in this book detailing the proper members and the proper manner in which to install those products—to get the results you're paying for.

What Is a Member?

The word *member* is an all-inclusive term, which can apply to any part of any assembly. For example, a *stud* is a member used to frame a wall. A *joist* is a member used to frame a ceiling or a floor. You will see in this book when I mention 1x_ and 2x _ members, which refers to a nominal-sized member, but lets you know that the remaining dimension doesn't really matter to make the point in that particular case. So the 1x could be a 1x4,

1x6, 1x8 (etc.), and the 2x could be 2x4, 2x6, 2x8 (etc.). Another thing to note is how the terms 1x and 2x are used in general. In the case of lumber, the dimensions noted are referred to as nominal sizes, which relates to the rough-sawn dimensions of the lumber. Thus a 1x4 measures 1"x4" when it is rough cut, but when fully dressed it measures 3/4"x3 1/2" (sometimes even a little less than that depending on the sawmill). After the publication of the first edition, I received a lot more questions regarding these terms than I thought I would. For example, people asked if it would be OK for them to use 3/4" lumber because they couldn't find 1" lumber. They wondered if it would work the same. The answer is that yes, it will work just fine because we are talking about the same materials. Please understand that this is not a construction manual. If you don't understand construction terminology, it would be worth your while to buy a book on that topic.

Another point I want to make is this: If you don't find a recommendation for a particular product in this book, then it's probably because it's not the most cost-effective way to go.

There are a lot of isolation products on the market that will "get the job done," but when you set them down side by side with standard construction and do a square foot analysis on the isolation received versus using standard construction with drywall (as one example), there is no way that they can compare.

It isn't that their claims of isolation value are false—they aren't—it's just that they are so expensive that they cannot compete with standard materials. Remember that the basics are mass, mass, and more mass, and some of the cheapest mass you can buy is standard 5/8" fire-code drywall.

The last thing I want to touch on is the topic of quality control.

Ensuring Quality Control Protection

In the construction industry, we typically have a project manager who buys out the project and makes the administrative decisions required to bring a project to successful completion. We also have a project superintendent who manages the day-to-day scheduling and on-site supervision necessary to verify that the work has been completed and is installed correctly.

Doing the Job Yourself

The project superintendent is the on-site quality control inspector. It is his responsibility to make certain that everything is done according to plans and specifications through each step of the project. In the case of your home studio, you are probably going to be taking on that role.

If you want to do all of the work yourself, make it easier by taking it one step at a time and referring back to the book as you move along. But, if you hire contractors to do the work for you, it will be much harder for you to follow the progression of the work between your daily job and the performance of their contracts. Work can easily get buried before you ever have a chance to see it, and this is a recipe for disaster if your contractor decides to "cut a few corners."

Remember that what is outlined in this book does not reflect standard construction practices for 99 percent of the work that these contractors will ever perform, so they might get the attitude that it is "overkill" and not really required in order for them to get the job done "right." However, what's outlined in this book is what's really required to attain the extra degree of isolation necessary for a studio environment.

Hiring the Job Out

Basically, it's up to you to do one of two things to protect yourself. The first possibility is to find someone who knows construction, for example, a trusted friend. Spend some time going over the requirements in this book with them, making certain that they understand exactly what needs to be done. Then have them do inspections for you along the way to ensure that the contractor is doing the work properly.

The other possibility is for you to specify in your contract with the subcontractor some work limits that will allow you to perform the proper inspections.

For example, if the contractor could install two layers of drywall in one day, he might decide that the caulking of the first layer is overkill and not bother to do it at all. You would never know this unless you began destructive testing of the work, such as removing the caulk and backer rod from the second layer to see the first layer. You could specify (in your contract documents) that the second (third, fourth, etc.) layer of drywall could not be installed prior to inspection of the completed previous layer.

If the contractor had someone to call for inspections (meaning you), he could schedule this inspection so that his work would move along seamlessly. However, if he has to wait for you to inspect at the end of each day, he will probably charge you a premium for this. Pay the premium; otherwise, you might get only part of what you are paying for.

The same goes for treatments behind the wall, prior to the subsequent sheet. Although you can count on your building inspector to ensure that the electrical wiring is done correctly, the inspector will only insist on adherence to the code and nothing beyond that. Thus, you could pay for certain things and not have them in the end. In certain areas around the world, you might not even have building officials overseeing the work that is taking place.

So, if you are going to contract out part (or all) of the work on your studio, make absolutely certain that you outline all of the steps where inspections will be required. Insert clauses requiring removal of materials (at no cost to you) to allow for inspections of any work that is covered up prior to your inspection (*and acceptance*) of the work in place.

Trust me when I tell you that you will pay extra to do work as outlined in this book, that is, compared to standard construction techniques, but if you follow the techniques outlined here, you will achieve your goals for creating a home studio you can be proud of.

Endnotes

[1] Photography by Author, printed with permission of Tony Bongiovi and Sonalysts Studios.

[2] Photography by Dennis Cham, printed with permission of Denis Cham and Hit Productions.

[3] Photography by Max Dearing, printed with permission of Max Dearing and Dark Pine Studios.

CHAPTER 2

Modes, Nodes, and Other Terms of Confusion

Understanding how sound works (within an enclosed space) helps determine how to design recording studios properly. You pick up the best speakers you can afford—special amplifiers—all made specifically for recording studio environments, all designed to avoid adding color to your mix. Then you place them inside a room that colors the sound anyway. Room modes can lead to uneven levels in frequency and longer sound decays than normal. In a critical listening room, like the control room in your studio, you need to hear exactly what's coming out of those speakers. You don't want to have the levels affected by anything other than how you set them on the board. But the room you're in can do just that—make some notes louder, some softer, and some go away completely. Understanding what causes this discrepancy, and how to avoid it (or at least minimize the problem), is an important part of designing a room in which you will enjoy working. Let's look at how all of this works together.

All frequencies act in waves; however, lower frequencies tend to be less directional than higher frequencies. If you think about your setup on a stage, you'll see that this is true. No matter where you are on the stage, including sitting behind a set of drums, you can feel the bass even if you don't hear it. That's because the bass frequencies get into the stage (the building), and they envelop everything around them.

But unless you have monitors, you won't be able to hear the vocals, the keyboard, or the lead guitar. The drummer especially suffers from this problem. Let's look at how this affects you in a room and take a brief look at sound itself.

Sound

Music is made up of a series of tones (frequencies) that contain amplitude (volume). There is a third variable, velocity (the speed of sound), but for the purposes of small rooms and this book, we won't worry about calculating this. I say "calculating" because the speed of sound is not a constant thing, it changes (quite drastically) with variations in temperature, humidity, barometric pressure, and elevation, all of which are conditions that exist in relation to your room. In this book, we will just use the conventional speed of 1,130fps (feet per second) for any calculations we might make.

Amplitude

Amplitude refers to the difference between the maximum and minimum pressures within a soundwave. Pressure fluctuations in these waves are symmetrical about the current atmospheric pressure. In order to keep things simple, we will use a reference value of zero and show maximum pressure as a positive value above zero and the minimum pressure as a negative value (see Figure 2.1).

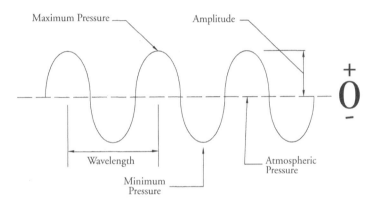

Figure 2.1
A typical soundwave.

Frequency

Frequency refers to the number of waves (pressure peaks) that travel a distance in one second. For example, low E on a four-string bass is 41.2Hz (based on A440 tuning) and thus has 41.2 waves traveling 1,130 feet in one second, while a 1kHz (1,000Hz) sound would have 1,000 waves traveling at 1,130 feet in that same second.

Wavelength

Wavelength refers to the physical distance between successive pressure maxima and is thus dependent on the speed of sound in the medium divided by the frequency of the wave. The equation for this is:

$\lambda = V/f$ [1]

Where:

λ = wavelength (feet)

V = velocity (speed of sound)

f = frequency (Hz)

Thus, the 41.2Hz frequency has a wavelength (distance between peaks) of 27.43 feet (8.37m) rounded to two decimal places 1,130/41.2 = 27.427, while a 1kHz frequency has a wavelength of 1.13 feet (0.344m) 1,130/1,000 = 1.13.

Room Modes

Every room, including even the most professional studio, has what are called "room modes." Let's begin with an understanding of what our real concern is here, which is frequencies from 200Hz and below. (I know Everest[2] says 300, but I think 200 is a more reasonable cutoff, since this is roughly G below middle C—certainly a midrange note in my book—and frequencies above this can easily be absorbed with fiberglass.) For now, let's walk through what creates room problems acoustically. We'll deal with how to treat them in Chapter 10, "Putting It All Together."

A room mode is created by a room dimension coinciding with a particular frequency's distance of travel. This sounds quite simple, but in reality gets a little more involved if you are really going to understand it. All rooms have a unique distribution of low frequency standing waves, also known as *modes*. Modes are resonances of a room governed directly by the dimensions of a room. The math for figuring out the modes of a room is[3]:

$$f = \frac{V}{2}\sqrt{\left[\left(\frac{n^x}{L_x}\right)^2 + \left(\frac{n^y}{L_y}\right)^2 + \left(\frac{n^z}{L_z}\right)^2\right]}$$

Where:

f = resonant frequency in Hz

V = 1130 ft/sec or 344 m/s (speed of sound in air)

nx, ny, nz = room mode numbers $\geq$ 0

Lx, Ly, Lz = room dimensions in ft or m

The reason you need to know this is due to what can happen when room dimensions are the same as wavelengths (also 1/2 wavelengths, 1/4 wavelengths, etc.). As mentioned previously, one of the goals of recording is to capture the song being played as truly as you possibly can. Once this is completed, you want to mix the final product down, while hearing the mix reflect exactly what is coming out of the speakers on your system (hopefully, an accurate reproduction of what you recorded). Room modes can affect the amplitude (volume) (see Figure 2.1) of a note by increasing or decreasing it.

Modal Waves

This phenomenon occurs when a room dimension coincides with the wavelength and the distance of travel is not great enough to allow the wave to decay naturally to the point that the incidence of it crossing itself (head-on) is neither constructive nor destructive.

A constructive action boosts the amplitude of a wave, while a destructive action reduces that amplitude. Destructive areas within a wave are referred to as *nodes*. Nodes are areas where the sound decreases in amplitude. Constructive areas are referred to as *antinodes*. Antinodes are found between nodes and are the locations where the sound increases in amplitude. All of this becomes more problematic the smaller and more reverberant the room is. A room that is dead (anechoic) will suffer this to a lesser degree. However, due to the fact that it will have no natural reverb or life within it, it will end up being a very boring room.

When a tangential mode is excited in a room, the resonant standing wave is "set up" and the resulting pressure distribution will be similar to that shown in Figure 2.2. The dark gray shades in the contour in Figure 2.2 are areas where sound pressure is at a maximum. The lighter areas in the "troughs" of the contour are areas of minimum pressure. This contour would correlate directly to what is heard when this particular room is excited by a frequency near the {3,1,0} tangential mode. In the case of this room, the dimensions are roughly 6m x 4m x 2.8m, and the frequency in question is roughly 96Hz. This response is dependent on the location of the loudspeakers. The loudspeakers for this model were placed very near a wall. If the speakers were located in or near the "trough" areas (nodes) instead, the darker areas (antinodes) would not be nearly as "hot" with sound energy because the loudspeaker would be unable to excite this particular mode. Likewise, if listeners were seated in or near the "trough" areas, their perception would be that the bass in the 96Hz range was "thin" or somehow lacking. The frequency in Figure 2.2 roughly corresponds to G2 on the musical scale, or G an octave and a half below middle C. Mixing songs where this note is played frequently by a guitarist or bassist (in a room with these dimensions) may sound "thin." Not knowing that this is a room problem, the person mixing may overcompensate by boosting the low-frequency EQ in this range. Consequently, the mix would not translate well, and it would probably sound "muddy" or even distorted on the low end when played back on another system.

Let's back up and look at our 41.2Hz example for a moment. The wavelength is 27.43. If the room length, width, or height is that same dimension, the sound pressure within the room will create a standing wave. If the listener were walking around the room, there would be a series of peaks (antinodes) and nulls (nodes), which would correspond to the wave itself. Depending on the position of the listener in relation to the wavelength, he might hear a peak, null or normal amplitude in the tone.

Figure 2.2
Tangential modal wave in excited state.

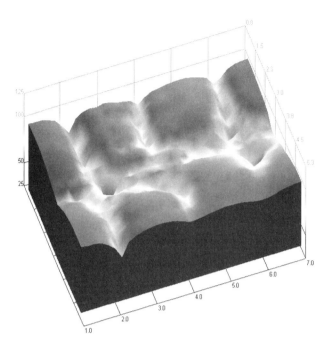

Where this begins to get a little more involved (like we needed that) is that frequencies that are harmonics of a prime frequency related to a room dimension can also be excited. Any whole number multiple of the frequency will also be a problem. So if 55Hz is a problem, then 110 is a problem, 165, 220, 275, etc. This corresponds to 1/2 of the room length, 1/3 of the room length, 1/4 of the room length, 1/5 of the room length, and so on.

By the same token, constructing a room with a dimension of 13.72' (1/2 of the room dimension above) will provide exactly the right conditions for all of those frequencies to still get excited, causing the same problem.

If you haven't thought about all of this before now, you're probably wondering why any of this matters to you. The answer is that standing waves, or any wave that acts like a standing wave (more on that in a moment), can affect the amount of a particular frequency that you add (or subtract) during mixdown, due to the fact that it sounds louder or softer than it really is in the mix. It can also cause various musical parts of an instrument being played to vary in amplitude, if you used a room mic for ambiance.

One of the ways of attempting to deal with this problem when recording in small rooms is to close mic the instrument or amplifier. When doing this, you can lower instrument volumes, which at least helps to take some of the room out of the equation. However, if your mic sits in a null, nothing can help you capture a note that isn't heard. However, in the control room, you do not have that option. In the control room, you generally want to lie back, set your mix levels, and generate the sound you would hear if you took this to another room to play it back. You could always wear headphones (to take the control room out of the equation), but this is not the preferred method of mixing down a recording.

On the issue of headphone use: Although you can use them, they become awkward to work with for any real length of time. Also, they do not really represent a listening area. By this I mean that a good listening room brings a sense of spaciousness into the listening experience, and even the average listening room creates a feel that cannot be duplicated with headphones. It is, however, not necessarily a bad idea to check your mix with them occasionally. It gives you another tool in your arsenal to see how your mix will translate in the real world, but that is a world of difference from working with them exclusively.

Suppose for a moment that your listening position is in a location where that same E (41.2Hz) we've been discussing here has a peak of 6dB. (We'll look at exactly how we get this 6dB peak in a moment.) Because the room is artificially amplifying this sound in your mixing spot, you would naturally lower the levels until the mix sounded right. But when you walked out of that room and began to play this in another location—one that this phenomenon did not exist in—you would suddenly find that your mix had holes in it where the bass belonged. For example, a 6dB cut would sound like the bottom just fell out. In reverse, should you be sitting in a null, the mix would end up being bass heavy because you would compensate for it by boosting that frequency. The design goal for a mixing room is to be able to hear exactly what's coming out of your speakers without the room affecting the outcome.

Now to understand this all in greater depth, compare a mode to a series of colliding balls, also known as *Newton's Cradle* (see Figure 2.3). We've all seen this one time or another in various forms.

It's a case of an outer ball (or multiple of balls) being swung so that it impacts an adjacent ball—with the result being an equal number of balls on the opposite side being propelled at (apparently) the same speed and distance—with this repeating from side-to-side once set into motion. The balls in the middle never seem to move. They are actually being placed in a state of compression, transferring this energy to an adjacent ball, at which point they are in a state of rest until they are again compressed and do the same in reverse.

Figure 2.3
Here is an example of Newton's Cradle.

Now picture a wall sitting on either side of those central balls—with the walls moving in and out (to impart the energy), instead of having those outside balls in motion. To the viewer (in either case), those balls in the middle don't move.

If you could physically see the excited mode of a room, you would be amazed to see virtually the same thing. The wave doesn't move—it just sits there and grows in level.

In the case of the balls, you have impact, compression, transfer of energy, and then a period of rest. In the case of the room mode, you have a spring effect, thus you have compression, tension, compression, tension, with no point of rest.

With sound then, you have a source point sending a soundwave, which strikes a boundary, and the signal is returned toward the source. While this is taking place—at the exact point in time as the original signal is being returned—a second wave is being sent from the original source location. They meet in the middle of the room and pass one another. We now have two waves passing in opposite directions from one another—perfectly in sync. This is the point where the wave becomes standing. The points of pressure and velocity become essentially locked in place and do not move. They simply begin to build within the space.

At the initial intersection of one another, the amplitude is increased or decreased by 6dB, depending on the particular location of the listener within the space. But if the duration of the tone generated is long enough, this can grow quite dramatically. It isn't unusual during room testing to see peaks and dips greater than 20dB.

OK, before you say anything, let me clarify. "*This would be the case if the speaker were perfectly flush with the face of the wall. But if it isn't, how would this still work?*" (You were thinking that, weren't you?) Remember when I said earlier that bass frequencies traveled in waves? I explained that they permeate that which is around them. For example, they build up within a space similar to the way waves spread out in a pond when you throw a rock into the water. So, even if the speakers aren't at the edge of the room, soundwaves can still create a standing wave by exciting a room mode. It doesn't matter whether the completion of the cycle occurs at the edge of the room or 1' from that edge. When a soundwave meets itself head on, it will create a standing wave. I do want to point out to you that a mode is able to be excited to a greater extent from the room boundaries, and because of this, you can minimize the level of this phenomenon by proper speaker placement. However, regardless of the source of a wave, if its dimension relates to a room dimension, it can generate a standing wave.

Non-Modal Waves

Although technically only a room mode creates a standing wave, I like to use the term *non-modal waves* to describe a phenomenon that is similar to a room mode but has different causes. Whereas a standing wave is related directly to room dimensions, non-modal waves can be related to the distance of a speaker to a surface, an instrument to a surface, and so on. Let's look at a few examples of non-modal standing waves.

Back Wall Interference

Back wall interference (reflections off the wall behind the listener) is probably as significant an issue as modal standing waves. In this case, where the listener is located, between the speaker and the interfering boundary, becomes important. As the listener moves forward and backward, he will hear peaks/dips at different frequencies. In this case, as the listener moves forward in the room, the effect of this interference on the various frequencies will become less apparent. When we look at room design, we will examine speaker placement and seating positions to minimize this problem.

So, now we have room modes, nodes, antinodes, non-modal standing waves, speaker placement, listener location—and all of these are problematic issues to contend with. Could it get any more challenging?

Yes, it could, and it does. In addition to all of this, you can have nulls created by something called "speaker boundary interference response."

SBIR

SBIR (speaker boundary interference response) is the coupling of the listener and the loudspeakers to reflections from surfaces adjacent to the speaker. (The speaker is sitting between the boundary and the listener, in this case.) This problem again exists primarily with (but is not necessarily limited to) low frequencies. It is a combination of the interference (both constructive and destructive) from the loudspeaker's direct sound and reflections from the room boundaries that can cause severe peaks and nulls. When a sound is transmitted from a speaker, the waves reflect off a nearby surface and the reflected wave is coupled with the direct wave. Treatments between the speaker and the problem surface can generally deal with this.

Other Reflective Problems

You also have reflected sounds that can destroy the stereo imagery of your recording. Although these effects are easier to deal with than the modal issues that plague small rooms, they will have to be dealt with nonetheless.

Flutter Echo

Flutter echo is a distinctive ringing sound caused by echoes bouncing back and forth between hard, parallel surfaces, following a percussive sound, such as a handclap.

To minimize flutter echoes, which can plague even a studio having a perfect RT60 across the band, certain precautions can be taken. If you're building from scratch, facing walls can be constructed out of parallel by at least 1 in 10 (6%), but if this isn't possible, some form of mid/high absorber can be applied to one or both walls to reduce the problem. Figure 2.4 indicates causes of flutter echo.

What Is RT60?

T60 (also known as RT60) is reverberation (in a space) measured as the time required for a reduction of sound pressure levels by 60dB.

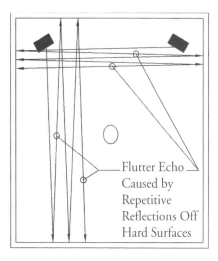

Figure 2.4
Flutter echo.

Comb Filtering

Another form of acoustic distortion introduced by room reflections is comb filtering. Comb filtering occurs due to interferences that exist between a direct sound and the reflected sound from the same source. In control rooms, you are primarily concerned with the interaction between direct sounds and their first-order (i.e., single-bounce) reflections. These reflections cause time delays because the reflected path length between your ear and the source is longer than the direct sound path. Thus, when the direct sound is combined with the reflected sound, you'll experience notching and peaks, referred to as *comb filtering*. The term "comb filtering" comes from the plot that this data provides, which resembles a comb. Figure 2.5 indicates the cause of comb filtering. Figure 2.6 is a graphical representation of the comb-filtering effect.

Figure 2.5
The cause of comb filtering.

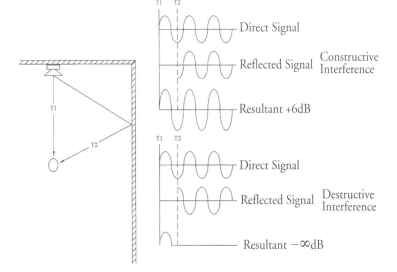

Figure 2.6 Comb filter plot indicating peaks and pips in amplitude.

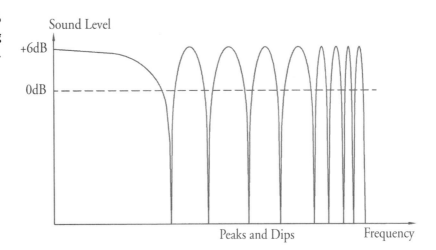

Early Reflections and Stereo Imaging

You can also have the stereo image destroyed by early reflections from the wrong speaker. This takes place when a sound from one speaker bounces off a sidewall and enters the ear a split second after the direct sound from that ear's speaker. Because the delay is only a split second, the mind does not differentiate between the two sources. It tricks the brain into believing the sound came from the speaker located on that side of the room. This occurrence destroys proper stereo imaging. For example, your left ear believes that it hears (from the left speaker) sounds that are actually emanating from the right speaker, which makes it impossible for you to properly mix a true stereo image of your recording. Figure 2.7 shows you the cause of this early reflection issue.

Figure 2.7 Early reflection destroying a stereo image.

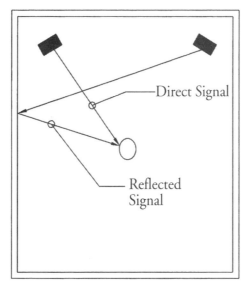

Because of the fact that this problem exists within the mid- to high-frequency range, it is another quirk that can be handled fairly easy. You will treat this by creating a reflection-free zone within the room.

Mode Analysis

"Rod, I used one of the 'best' room modes to design my room, I paid attention to detail, everything is constructed, and I ran a test on my room, but the modes are not what were predicted with mode analysis. What did I do wrong?"

It is very possible that you could design a room properly, completely test it after construction, never make a mistake, and yet have different results with modal behavior than you expected, based on an analysis with a room mode calculator.

I have never seen this situation completely explained anywhere, but when I think about it, at least one contributing factor comes to mind. Keep in mind that this is my *own* conclusion based on my understanding of this subject, and it has not been proven by any tests I know of to date. But the conclusion is perfectly logical to me, so I offer it for your consideration. At some point in time, I am going to have to develop a construction model for this and have it tested.

I believe that part of the reason for this is due to the fact that a modal analysis of any space makes certain assumptions of that space. One is that the room is rectangular in shape. Another is that the modal boundaries are an infinitely rigid dense material that sound is incapable of escaping from. Although the math works from the perspective of calculating the effects (if these infinitely dense walls really existed in your room) in the vast majority of cases, this is not anywhere near reality, especially with small home studios.

We all construct walls from which sound escapes. In fact, we might build a wall that has a center frequency, which allows sound at that frequency to strike a boundary directly on the other side of the wall and then re-enter the room through that same wall.

Let's take a look at a basement studio for a moment. Suppose that you have a basement with a wall-to-wall dimension of 36' 7 3/8". Within this space, you construct a "room within a room" studio, with the walls 1" clear from the face of foundation wall, 2x4 construction with drywall over that, and when completed the walls have a center frequency of 30.87Hz (which just happens to coincide with the low B string on a five-string bass guitar). In this case, a note from that B string would pass through the wall "like a hot knife going through butter" and be reflected off the outer boundary back into the room. Because your concrete walls are 36' 7 3/8" from one another, which coincides with the wavelength of that B string, you could have a room mode you can't account for if you base the mode analysis strictly on the room itself.

Now understand, I am not suggesting that you ignore room modes during design, nor am I suggesting that an analysis of room modes never makes sense. Nothing could be farther from the truth. I am simply explaining that there is much more to this than analysis, math, and acoustic theories because when it comes to the predictive models in the acoustic world, things once constructed do not always behave exactly as predicted. Testing after the fact is always required to verify the treatments you will require within the space.

This is one of the reasons that many thousands of dollars are spent testing isolating assemblies, room treatments, etc. The math involved is a good thing because it gives you a fairly good idea of whether you're headed in the right direction. However, the only way you can ever know with any degree of certainty whether your predictions were correct or not is to perform tests. So begin with the best you can, build it the best that can be built, and then test for actual conditions before you begin any room treatments.

Room Sizes

By the way, all of this becomes more problematic the smaller the room is. Modal activity in relatively large rooms might be small in nature due to the natural decay rate of waves. As rooms become smaller, the amplitude of the wave has less decay time during its travel and thus exhibits greater constructive/destructive interference. Let's look at these rooms a little more closely.

Large Rooms

For home studios, I would describe large rooms as rooms with minimum lengths of around 30', widths of around 20', and average room heights of not less than 12'—or combinations thereof equaling greater than 5,000c.f. (cubic feet). In large rooms, there is generally enough travel distance so that sound has a chance to decay to the point that constructive/destructive actions in the room will begin to be minimized. What effect there is, ends up being fairly easy to manage from a treatment point of view. In addition, large rooms have the advantage of being able to institute designs, including splayed walls, which, if the splay is large enough, can create a modal zone that spans multiple frequencies in any given wall dimension. You can construct a room with acoustic qualities rivaling most professional studios if you can provide this much volume in your room.

Medium Rooms

Medium-sized rooms—and by this I mean rooms of 2,000 to 5,000c.f. in size—tend to exhibit more modal activities, but are generally large enough that some care can be taken to alleviate problems relating to modes without having to go to extremes.

Small Rooms

Small rooms, however, (rooms less than 2,000c.f. in size) become more and more problematic as you decrease the room area. And in some small rooms, there can be dimensions that cause even more problems than you might expect.

Occasionally, I have people come and ask how to deal with rooms that are cubed dimensionally, for example, 8x8x8 or 10^3. I tell them that these rooms are to be avoided like the plague. Not only are they small in area (512c.f. to 1,000c.f.), but they also have exactly the same modes in all three dimensions—a modal nightmare that will drive you crazy. The only advice I can give them if they can't make physical changes to the room dimensions is to either find a new room or make the room completely dead.

You should know that there are also problems inherent with making rooms completely dead. The first is a matter of human comfort. We are not accustomed to existing in a space that is without any amount of reverb, and thus we tend to tire quite quickly when confronted with working in this environment. The next problem is that one tends to overcompensate for the lack of natural reverb by adding too much artificial reverb into the mix. Having said that, I would like to clarify one thing: you can learn to compensate for a lack of natural reverb.

You can learn to compensate for a mode, non-modal standing waves, and any other acoustic anomaly that might exist in a room. If you truly pay attention to what is going on in your mix and learn your room, you can make the adjustments necessary to mix right, even if your room is wrong. It's been done before, and some real great music has come out of some really bad rooms. But, if you do this as a hobby—for example, if you don't do this day in and day out—you may never really "get it," and even if you do, it will still never sound "perfect" to you while you're in the room, even if it sounds great when you finally play it outside. Besides, that all sounds (to me) like a little bit too much work, and I don't know about you, but I do this for fun. It's better if we can make rooms that minimize these problems. (Even if I did it for a living, I would want to enjoy the experience.)

Types of Room Modes

All of our work so far has been to focus on sound moving side-to-side, front-to-back, or top-to-bottom. However, this is only one type of room mode. There are actually three types of room modes: axial modes, tangential modes, and oblique modes.

Axial Modes

The first mode is the axial mode, which is exactly the concept we were discussing earlier. This mode is based on the simple dimensions of walls opposite one another and the dimension of the floor to ceiling.

Take a look at Figure 2.8, which is an example of axial modes in a simple rectangular room. The mode is created by reflections between two opposite surfaces. Understand that this is not an either/or situation. You will have modal activity from all of these surfaces. That's why it's so important to make certain that your room ratios are such that they reinforce modes in multiple directions.

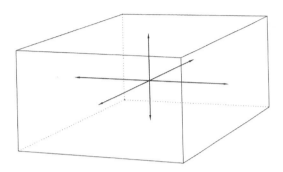

Figure 2.8
Axial modes in a room.

Tangential Modes

Tangential modes are formed by the relationship of any four surfaces within a room. When the travel distance formed by the four surfaces coincides with a frequency wavelength, the tangential mode can become excited. Tangential modes require two times the power of an axial mode to produce the same change in amplitude (pressure level). In Figure 2.9, you can see an example of a tangential mode in a rectangular room. In this case, you see one modal activity in play (for clarity's sake), which includes only the four walls, but it could just as easily be any two walls with the ceiling and floor in play as well.

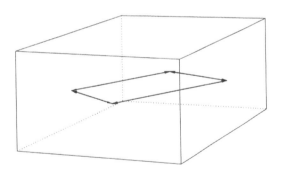

Figure 2.9
Tangential modes in a room.

Oblique Modes

Oblique modes are formed by the relationship of all six surfaces within the room. When a travel distance formed by these six surfaces coincides with a frequency wavelength, the oblique mode can become excited. Oblique modes require four times the power of an axial mode to produce the same change in amplitude (sound pressure).

Figure 2.10 indicates the sound travel required to produce an oblique mode. As shown, this mode becomes excited when the travel of the sound takes all six surfaces into account. Because of the power requirements of this mode in relation to an axial mode, this will generally create much less interference—even in small rooms—than either axial or tangential modes.

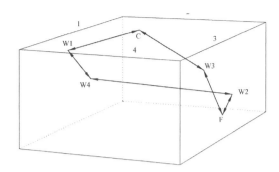

Figure 2.10
Oblique modes in a room.

Figures 2.8 through 2.10 are simplified, to say the least. They are intended to help you understand the manner in which these modes work, but they are also slightly deceptive because they paint a picture of a very tight soundwave. A laser beam, with properly placed mirrors, would travel as shown above, but soundwaves are not that directional, especially low-frequency (LF) soundwaves. As discussed in the beginning of this chapter, LF waves are very broad and tend to permeate the room. Thus, although ray tracing (as shown above) can help to understand how this works, in general, it is not really representative of LF modal behavior.

In properly designed large- and medium-sized rooms, I do not become overly concerned with oblique and tangential modes. Because mixing is generally done around 85dB (this is about the maximum sound level that can be used without tiring one's self out quickly), the decay rate of the sound should allow for very little amplitude interference. This is due to the distance of travel within the space. Treatment of any interference that exists is relatively easy to handle. But small rooms are a different story. In small rooms, I would recommend that any modal analysis include all three modes.

For the purposes of small room design, there have been numerous studies of the effects that room ratios have in relation to modal activity. Next, we will look at some of the outcomes of those studies.

Room Ratios

In order to design rooms in which these types of problems can be minimized, a series of ratios have been established over the years specifically for small room designs.

Some of the more popular ratios are as follows:

(In all cases, 1 is equal to ceiling height, and walls are represented as a ratio relating to the ceiling height; thus, Height × Width × Length.)

	H	W	L
L. W. Sepmeyer:[4]	1:	1.14:	1.39
	1:	1.28:	1.54
	1:	1.60:	2.33
M. M. Louden:[5]	1:	1.40:	1.90
	1:	1.30:	1.90
	1:	1.50:	2.10

Please note that the orders listed above are (top to bottom) the first to third best ratios in both cases. There are many other ratios that can be found, but the listed "first best" by both Sepmeyer and Louden are two of the more widely used in professional studios worldwide.

Mode Calculators

I want to touch on the subject of mode calculators for a moment. There are literally hundreds of these on the Internet. Some are much better than others. Personally, I don't use them. The reality is (especially in small rooms) that in the end you are going to have to treat your room to make it right. If you use known room ratios to begin with, then you are that much ahead of the game. If you do not have the space to do that—for example, if you have such a small space that you just can't afford to give up any real-estate to make these adjustments—then just build your room and deal with what treatments you have to in the end. Although running a mode calculator is easy, understanding it is altogether different. You could be building your room for a long time before you figure out what it's about.

However, should you decide to get into it and calculate room modes beforehand, don't get nervous if you use the ratios listed here (or some other of the famous room ratios) when you check your room modes. Because when you do, you'll see a lot of excited modes. This is to be expected for small rooms. One of the things that makes these rooms work is that when you have enough modes excited, things tend to smooth out. If you could create a room (for example) in which all frequencies were modes—and all were excited equally—then for all intents and purpose, the modes would not exist.

To make it easier for you to identify frequencies with distance, I have included a frequency chart in the following pages.

Frequency Chart

These tables were created using A4 = 440Hz

Speed of sound = 1130ft/s

("Middle C" is C_4)

Note	Frequency (Hz)	Wavelength (cm)	Wavelength (ft)
C0	16.35	2100	69.11
C#0/Db0	17.32	1990	65.24
D0	18.35	1870	61.58
D#0/Eb0	19.45	1770	58.10
E0	20.6	1670	54.85
F0	21.83	1580	51.76
F#0/Gb0	23.12	1490	48.88
G0	24.5	1400	46.12
G#0/Ab0	25.96	1320	43.53
A0	27.5	1250	41.09
A#0/Bb0	29.14	1180	38.78
B0	30.87	1110	36.61
C1	32.7	1050	34.56
C#1/Db1	34.65	996	32.61
D1	36.71	940	30.78
D#1/Eb1	38.89	887	29.06
E1	41.2	837	27.43
F1	43.65	790	25.89
F#1/Gb1	46.25	746	24.43
G1	49	704	23.06
G#1/Ab1	51.91	665	21.77
A1	55	627	20.55
A#1/Bb1	58.27	592	19.39
B1	61.74	559	18.30
C2	65.41	527	17.28
C#2/Db2	69.3	498	16.31
D2	73.42	470	15.39
D#2/Eb2	77.78	444	14.53
E2	82.41	419	13.71
F2	87.31	395	12.94
F#2/Gb2	92.5	373	12.22
G2	98	352	11.53
G#2/Ab2	103.83	332	10.88
A2	110	314	10.27
A#2/Bb2	116.54	296	9.70
B2	123.47	279	9.15
C3	130.81	264	8.64
C#3/Db3	138.59	249	8.15
D3	146.83	235	7.70
D#3/Eb3	155.56	222	7.26

Note	Frequency (Hz)	Wavelength (cm)	Wavelength (ft)
E3	164.81	209	6.86
F3	174.61	198	6.47
F#3/Gb3	185	186	6.11
G3	196	176	5.77
G#3/Ab3	207.65	166	5.44
A3	220	157	5.14
A#3/Bb3	233.08	148	4.85
B3	246.94	140	4.58
C4	261.63	132	4.32
C#4/Db4	277.18	124	4.08
D4	293.66	117	3.85
D#4/Eb4	311.13	111	3.63
E4	329.63	105	3.43
F4	349.23	98.8	3.24
F#4/Gb4	369.99	93.2	3.05
G4	392	88	2.88
G#4/Ab4	415.3	83.1	2.72
A4	440	78.4	2.57
A#4/Bb4	466.16	74	2.42
B4	493.88	69.9	2.29
C5	523.25	65.9	2.16
C#5/Db5	554.37	62.2	2.04
D5	587.33	58.7	1.92
D#5/Eb5	622.25	55.4	1.82
E5	659.26	52.3	1.71
F5	698.46	49.4	1.62
F#5/Gb5	739.99	46.6	1.53
G5	783.99	44	1.44
G#5/Ab5	830.61	41.5	1.36
A5	880	39.2	1.28
A#5/Bb5	932.33	37	1.21
B5	987.77	34.9	1.14
C6	1046.5	33	1.08
C#6/Db6	1108.73	31.1	1.02
D6	1174.66	29.4	0.96
D#6/Eb6	1244.51	27.7	0.91
E6	1318.51	26.2	0.86
F6	1396.91	24.7	0.81
F#6/Gb6	1479.98	23.3	0.76
G6	1567.98	22	0.72
G#6/Ab6	1661.22	20.8	0.68

Note	Frequency (Hz)	Wavelength (cm)	Wavelength (ft)
A6	1760	19.6	0.64
A#6/Bb6	1864.66	18.5	0.61
B6	1975.53	17.5	0.57
C7	2093	16.5	0.54
C#7/Db7	2217.46	15.6	0.51
D7	2349.32	14.7	0.48
D#7/Eb7	2489.02	13.9	0.45
E7	2637.02	13.1	0.43
F7	2793.83	12.3	0.40
F#7/Gb7	2959.96	11.7	0.38
G7	3135.96	11	0.36
G#7/Ab7	3322.44	10.4	0.34
A7	3520	9.8	0.32
A#7/Bb7	3729.31	9.3	0.30
B7	3951.07	8.7	0.29
C8	4186.01	8.2	0.27
C#8/Db8	4434.92	7.8	0.25
D8	4698.64	7.3	0.24
D#8/Eb8	4978.03	6.9	0.23

Now let's head into the next chapter and learn a bit about sound-isolating construction techniques.

Endnotes

[1] Source Unknown

[2] F. Alton Everest, (1909–2005), author of *The Master Handbook of Acoustics.*

[3] "A New Criterion for the Distribution of Normal Room Modes," Oscar Juan Bonello. *J. Audio Eng. Soc.*, Vol. 29, No. 9, September, 1981.

[4] "Computed Frequency and Angular Distribution of the Normal Modes of Vibration in Rectangular Rooms." *Journal of Acoustic Society of America,* Volume 37, No. 3, pages 413–423. (1965).

[5] "Dimension-ratios of Rectangular Rooms With Good Distribution of Eigentones." Acustica, Volume 24, pages 101–104. (1971).

CHAPTER 3

Isolation Techniques—Understanding the Concepts

First, please don't get too hung up on the term "STC ratings." STC (Sound Transmission Class) rating is a single-number rating used to compare different assemblies, based on the reduction in noise levels that the assembly provides. Partition sound transmission losses are measured by using ASTM E 90 "Standard Test Method for Laboratory Measurement of Airborne Sound Transmission Loss of Building Partitions and Elements" and are calculated using ASTM E 413 "Classification for Rating Sound Insulation."

The higher the STC rating is, the better the sound isolation value of the assembly will be. However, STC ratings were really designed and intended for the frequencies dealing with human speech and not for music. There is no standard (that I know of) that is specifically designed for recording studios. So although we will deal with STC for initially choosing an assembly to start with, we will leave it behind as we try to deal with isolating the frequencies below those that the STC ratings deal with.

Second, forget about building "soundproof" rooms—they don't exist. Any assembly will let sound through—it just depends on the volume and frequency of the sound. What you are really going to build are sound-isolating walls, ceilings, and floor assemblies, which affect the amount of sound able to travel through the assembly when completed. The challenge when designing a home studio is to figure out what you need to isolate yourself from (or whom you need to isolate from you), what the offending sound levels are, what the frequency ranges are, and just how much of that sound needs to be affected to get the job done right. For example, if you will only be playing an acoustic violin within the space, you live alone in the middle of the woods, 400' from your nearest neighbor, 500' from the road, and in a no-fly zone—then you can get away with standard construction techniques without worrying about outside noise bothering you (or you bothering anyone else). The only thing you will have to do is treat your space acoustically.

However, if you have neighbors 10' from your house, who play music outside in nice weather at 110dB, with the speakers pointed toward your room and they listen to rap music, then you have your work cut out for you to isolate yourself from that noise source.

Determining what you need to keep out (or "in"—once constructed, isolation works the same in both directions) is as simple as obtaining a sound-level meter and gathering data on the sound levels. Take some readings within the space you plan to use when the noise levels outside are at their worst. Alternatively (if you live in a quiet area), get some musicians inside your space and let them jam. Take some readings outside your house, both within 3' of the house and then near your property lines. Take some readings throughout the house. Make sure to take notes along the way, because this data will help you decide exactly how much isolation you need. It will also let you know where your areas of concern are within the house.

Sound-Level Meters

Sound-level meters are measuring devices for sound-pressure levels. These devices are as inexpensive as $40 for nonrecording meters to several thousands of dollars for professional data-gathering devices. Please note that the $40 meters generally do not read below 50dB, but if readings before any construction are below that level, you are in pretty good shape anyway. The Radio Shack, Digital-Display Sound-Level Meter (Model 33-2055) is a real good buy at $49.99.

If you don't have a comfort level with your ability to gather the data, you can always hire a local acoustician (acoustic engineer) to come over and take some readings for you. The overall cost of a few hours of readings (by a professional) should not cost you as much as a decent quality meter would.

We will discuss in more detail what you require, based on those readings, in Chapters 4, "Floor, Wall, and Ceiling Construction Details," and 5, "Window and Door Construction," when you learn about isolated assemblies. For now, let's focus on what you need to get the quality of the work that's going to be required to meet your needs.

For the purpose of constructing sound-isolating assemblies, certain conditions need to be met. Acousticians pay close attention to three areas when designing assemblies for isolation.

You need the following areas:

- ▶ Mass
- ▶ Airtight construction
- ▶ Flanking paths

Let's examine each of these items in detail.

Mass, Mass, and More Mass

Forget the myths you've heard over the Internet. Forget what your friends tell you. There aren't any magic beans you can buy that will stop your sound from bothering your family or neighbors. It takes airtight construction, decoupling from the structure, and mass to keep sound within (or out of) your room(s). Every doubling of mass increases isolation by an additional 6dB. This is known as the *Mass Law*. Although the Mass Law strictly applies to nonrigid (hence limp) assemblies, it can be used as an approximate guide to determine the amount of isolation possible, especially if you have a good idea what the wall design you're using is capable of before beginning construction.

Decibels

Let's take a moment to understand the decibel (dB) and how it works relating to the sounds you hear. The decibel is (strictly speaking) a way to compare power levels, where the two levels may be measured in watts, milliwatts, microwatts, or even kilowatts. To make it clear, picture a simple speaker cabinet containing one speaker. Assume an output of the speaker of 100 watts. Adding a second speaker adjacent to the first and set to the same 100-watt output will be an increase of 3dB (a doubling of power), which is an increase of 50% in amplitude. In order to take the next step, you will have to place two more speakers with the first pair. Now you have four speakers, all with 100 watts of output (four times the power) and an increase of 6dB in amplitude from the original signal. So 6dB is a doubling of SPL (sound-pressure level). However, the human ear does not perceive this as double in volume. The next step is to add four more speakers to the stack. A total of eight speakers, with 800 watts of output, for a net increase of 9dB. Now you're almost there. The human ear hears a doubling in volume at roughly 10dB. For the purpose of discussion in this book, 6dB will be referenced as either double or one-half as loud.

Unfortunately, it also holds true that for roughly every octave drop in pitch, sound isolation is halved, and although sound isolation for higher frequencies is relatively easy to obtain, isolation for lower frequencies is much more difficult to achieve.

To truly understand this, you would need to look at (and understand) how mass, stiffness, and damping come together to affect an assembly. Plus, you need to understand that panel resonance is affected at lower frequencies and coincidence at higher frequencies.

For the purpose of this discussion, understand that there is essentially no difference between a wall, ceiling, or floor assembly (not all floors are ceilings). When you look at a wall, the perimeter of that wall creates the outline of a panel, and the construction of that panel is called an *assembly*. The same is true when you look at a ceiling or a floor.

The assemblies may be constructed differently, but the concepts we will be examining here are exactly the same.

At lower frequencies (roughly 10–20Hz), sound transmission tends to depend mostly on how stiff the assembly is. Lucky for us, this frequency range is well below that of the music we are dealing with, so it shouldn't to be a problem for most home studios.

As frequencies rise, resonance begins to control transmission. Each assembly essentially creates a panel that will have its own center frequency. This is called the *resonant frequency*, and it consists of a fundamental frequency (having the greatest effect), and multiples of this fundamental called *harmonics*. This occurs at roughly 50 to 100Hz.

Picture a tuned drumhead. Connect a guitar tuner to a microphone and set the mic over your drum. When you strike the head, you can see the frequency the head is tuned to. (This is easier if the snares are off by the way.) That is the resonant frequency of that drumhead. A wall, floor, or ceiling assembly is much the same as that drumhead. A resonant frequency will pass through a wall like a hot knife through butter. For this reason, it would be a fairly perfect world if we could create assemblies with resonant frequencies below 10Hz. (Remember that we do not often use that frequency in most music.) The calculation for the resonant frequency of a wall is as follows:

$Fr = 0.45 * vl * b((1/1)^2 + (1/h)^2)$

and

$vl = \sqrt{(E/(p*(1-s^2)))}$

Where:

b = the panel thickness (m),

l and h = length and height (m), and

vl = the longitudinal velocity of sound in the partition (m/s).

In the calculation of vl:

E = Young's modulus of elasticity,

s = it's Poisson ratio, and

p = density (kg/m^3).

You can calculate harmonic frequencies for that same assembly by simply replacing the number 1 in the first equation with the required harmonic number. Frequencies between the resonant frequency and coincidence are the ones controlled most by Mass Law (roughly 100Hz to 1,000Hz). Those are the frequencies that we can handle strictly by adding mass to our system. While doing that, again remember that the doubling of mass increases isolation by (roughly) 6dB. This is the Mass Law.

That leaves us with understanding coincidence. For every frequency above a certain critical frequency, there is an angle of incidence for which the wavelength of a bending wave can become equal to the wavelength of an impacting sound. This condition is known as *coincidence.*

$Fc = v^2/(1.8 * h * vl * \sin^2(a))$

Where:

Fc = Frequency of coincidence,

V = the speed of sound in air (m/s),

h = the panel thickness (m),

vl = the longitudinal velocity of sound in the partition (m/s), and

a = the angle of incidence.

When coincidence occurs, it allows an easy transfer of sound from one side of a panel to the opposite side. This shows up as a big "dip" at the critical frequency. In thin materials like glass, coincidence frequencies begin somewhere between 1,000 and 4,000Hz. What's interesting to note is that speech frequencies begin in this same region. Above the critical frequency, stiffness begins to play an important role again. See how easy that was?

It's not really that difficult, because (in the end) unless you plan to become an acoustic engineer, understanding all the little details about how this works is not as important as understanding exactly how to construct your room. There is a belief among a lot of people posting on the Internet that constructing walls with materials that have different densities, or the same densities with a different thickness, will increase sound isolation.

I see people posting all the time that they have superior walls because they installed one layer of 1/2" gypsum board over one layer of 5/8" gypsum board over one layer of 5/8" medium-density fiber board. They believe that the wall is superior because each of the materials has a different TL value. Although I might understand their thought process, I cannot find any hard data to support that this is a better wall than it would be if they used the equivalent mass of standard 5/8" gypsum board. I always ask them how they tested their walls and what the actual TL value of the final product was at the various frequencies we're all interested in. The responses were pretty much always the same: "I didn't actually have it tested, but when I stand outside and the music is blaring inside, I can hardly hear it, so it must be better."

Actually, what they heard outside, or in another room of the house, does not indicate the value of the assembly over standard construction techniques, Only laboratory testing of an assembly will give you accurate answers. Without that, it's nothing more than guesswork. I will never suggest that you spend your hard-earned money on construction, when the outcome is not based on laboratory-tested assemblies. Rather, I would have you plan your construction based on the extensive testing that has been performed in certified laboratories.

Expand on that by simply adding mass to the wall to increase isolation. "Tested, tried, and true" is my motto, and I stand behind that because it's the only way I can assure you of the outcome you desire.

Airtight Construction

Simply put, where air goes, sound goes.

Some of the biggest problems I've seen over the years with sound transmissions from room to room have to do with a contractor's failure to pay attention to air paths created during construction between rooms. Contractors install walls designed (and tested) to provide STC ratings of 54, and when the job is finished, they have actual STC values as low as 34.

For example, think about the small cracks that exist at the bottom of walls formed between the bottom plate and the floor, which are generally ignored when finishing the rooms. A 1/16" crack located at the bottom of a wall 10' long is the equivalent of a hole through the wall 7 1/2" square in area. That's a hole in the wall 2" wide by 3 3/4" high. I am sure you can see why this would seriously reduce sound isolation room to room.

Similar weak spots exist with electrical boxes that penetrate the wall surface, the physical joints between the wallboard in the body of the wall in base sheets of drywall, as well as with the corners of the room where wallboards meet together. Yet these areas that I mention are the most commonly ignored items during construction, with no monies allocated in the construction budget for proper treatment. It is important that your contractors (or you if you intend for this to be a DIY project) understand and pay close attention to duplicating the exact construction details I outline in this book. Otherwise, you may as well throw your money to the wind—it will be an easier way to lose money. After all, construction is a lot of hard work to accomplish if you have nothing of value in hand when it's all done.

Eliminate Transmissions Through the Building Structure

This is another area that acousticians pay close attention to—and most people, including the "experts" doing the actual work in the construction industry, are not even aware it exists. Sound traveling through the structure of the building itself follows what are called *flanking paths*. An example of this can be seen if you take a wood stud, place your ear to one end, and have someone tap lightly on the other end with a hammer. You will hear the sound quite clearly through the stud itself—much more loudly than you would by just listening to that same tap in the room. Sound transmissions travel very well through the structure of a building, so disconnecting your studio from the rest of the building is very important if you want to isolate yourself from people walking around above you, the TV playing, or if your wife is trying to sleep while you're recording your band. Other sources of flanking sounds can be water pipes, drainage and sewer pipes, and ductwork for HVAC systems. Anything that interconnects different parts of the building with your studio space can create pathways for sound to travel through.

If you are in the process of designing a new building, care should be taken to design isolation into the structure. If your studio will be within an existing building, you can minimize the effects of flanking paths by isolating your room from the existing structure.

I'm not going to bury you in math, nor confuse you with the complex analysis that an acoustic engineer performs when working on room designs. If that's what you're looking for, there are already some excellent books on the market. I have no interest in duplicating their work or competing with them. Instead, I'm going to show you the practical side of this industry—how to achieve what you want most and give you the ability to make your music in peace. In the process, I am also going to let you know where not to waste your time and money. Sometimes, you just can't get from A to B, no matter how hard you may want to.

What to Avoid

It amazes me the number of times people have posted on sites that they have completed work on their studio and did not achieve the level of isolation they either expected or required. They ask how they can increase the level of isolation, to which I respond by asking them to explain exactly what they constructed, in as much detail as they can give. It usually takes me a while to get to the bottom of it, but I find that they had some "great ideas" to improve wall, floor, or ceiling assemblies. For example, they figured that they could gain more isolation by adding sheets of drywall within the cavity of a double-wall assembly. Their logic was: "Heck, if two sheets on each side were good, then putting two sheets in the middle must be better." Unfortunately, reality is just the opposite. Not only is it important that you have the right materials, but it is equally important that you put them in the right place. We'll look at what those "right places" are in the next chapter.

CHAPTER 4

Floor, Wall, and Ceiling Construction Details

Physically constructing a home studio isn't as simple as just buying a lot of wood and putting it together. There's a lot of thought and planning that has to go into this project in order for it to come out right. Deciding what you need and how to construct it in a manner that achieves your goal is probably more work than the actual work itself. In this chapter, you'll examine the options you have for various construction types, as well as the pros and cons of those techniques. By the end of this chapter, you should be well prepared to make the decisions you need to proceed with your design.

Floor Construction

In more cases than not in a home recording studio, you'll be trying to fit your room(s) within existing spaces. Hopefully, these spaces will be located on concrete slabs. I say that because it is much easier to achieve adequate isolation on concrete slabs than on elevated wooden decks. We'll spend a little bit of time looking at the "whys" of that later in this section.

Simple Concrete Slabs

Simple concrete slabs are what you would typically see in your basement or garage. They are generally 4 to 5" thick and probably don't have much in the way of reinforcing within the slab itself—that's if they have any reinforcing at all. The slab is poured above compacted earth, usually above a vapor barrier. It might even have some type of wire mesh in it, but this is intended to be at the top of the slab for shrinkage control, and does not really add any structural strength to the slab itself. Besides which, it is almost always placed badly (even for that purpose) due to the manner in which it is typically installed, which is lying flat on the ground, with the crew (who are placing the concrete) using their concrete rakes to pick it up into place. Not a good recipe to ensure consistent placement.

If a room within a room is constructed on an existing simple slab (as you might see in a garage with wood-framed exterior walls), sound can transmit through the slab and surrounding foundation itself into adjacent spaces. However, this sound transmission will probably be minimal compared to what loss you will have with ceiling and wall assemblies, depending on the building structure.

In the case of a full basement (walls buried below grade), the sound could also transmit through the slab into the foundation wall and then into the structure above. However, for the most part, earth is a fairly good damping material, and generally you can obtain the isolation you require by simply constructing your space above that existing slab.

Isolated Concrete Slabs

Isolated concrete slabs are those that are poured on earth, but only to the outside face of the walls for the room that sits on them. These are excellent slabs for studio design and that's how we constructed the slabs for Power Station New England. After pouring the main slab for the first room, a separate slab was poured adjacent to, but not touching, the first slab. The space between the two slabs was treated with an expansion joint material and then caulked to seal out moisture.

Figure 4.1 indicates typical details used for isolated slab construction. Note the use of the haunch at the slab edge. A *haunch* is a thickened section of concrete that is poured monolithically with a slab, in order to form what is effectively a footing capable of carrying a structural load.

This is created to provide adequate bearing for the room walls, which will sit on top of the slab in that location. There is a school of thought that a simple slab can be turned into an isolated slab simply by saw-cutting the slab in between the party walls of separate rooms. I don't suggest this method, due to the fact that a simple slab is not designed to carry a room load on its edge. It is very possible that the slab could crack and settle along the edge, causing problems with room finishes down the road. This is the sort of thing that you may not experience until after all the load (superimposed on the slab) is in place. It's a pain in the neck to finish your room, be fairly happy with your product, and suddenly have a wall on one corner settle $1/4"$ or more. The best scenario would be to remove the existing slab at least enough to install a proper haunch at the wall edge and then use a bonding agent to join the new to the old. If the existing slab has reinforcing in it, all the better, because you can remove the concrete without destroying the reinforcing and use it to help tie the new to the existing.

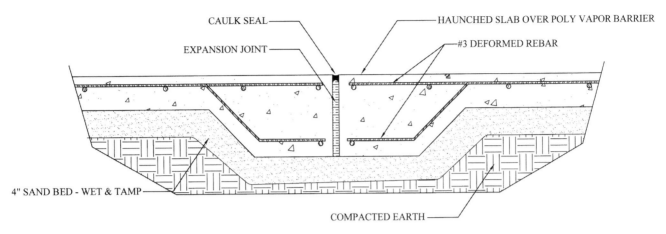

Figure 4.1   Here is a typical isolated slab.

Figure 4.2 indicates the details for adding bearing capacity to your existing slab. Note how the new slab/haunch pours slightly under the existing slab. This is referred to as underpinning, which helps to ensure that the two members act as one. You do not want the sections to act independently from one another.

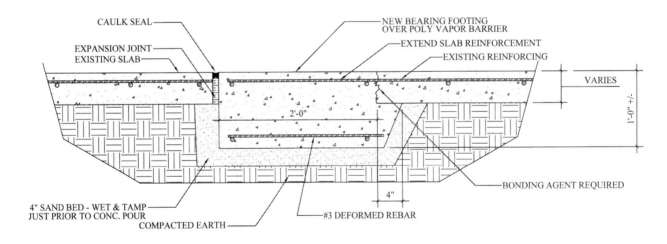

Figure 4.2   Here you can see added bearing at the slab.

By the way, just so you understand completely, this is not easy work. First, you really want to know if the slab has some reinforcing you can use to tie the old and new together. This requires the use of a jackhammer to open up a section of slab or special equipment to perform testing of the slab. Both pneumatic and electric jackhammers are readily available at all local equipment rental centers. The electric will not provide the same working speed as pneumatic, but it is easier to work with, due to the ease of running cable versus 1 1/4" hose and the tendency to weigh less. The testing equipment is generally not available for rental purposes and would require you to hire a consultant to test it or the purchase of the equipment, which is not inexpensive. Chances are that you will use the hammer. If you do find rebar or reinforcing wire installed within the slab, then the entire work must be done with that same jackhammer. As I mentioned earlier, it's a lot of work.

Locate the outside face of your room wall and snap a chalk line. At this point, you will saw-cut through the entire slab thickness. Use your jackhammer to carefully remove enough concrete to determine whether slab reinforcing exists. If it does, then locate a line about 2' inside of your room from the existing line. At this location, you will want to adjust your saw to cut only $1/2"$ deep into the concrete. This will provide a professional looking joint to finish your new concrete. Jackhammer out all the concrete into small pieces, which can be removed without removing the reinforcing. Save as much of the reinforcing as you possibly can. You will then have to bend that up along the face of the concrete in order to excavate for your new bearing point. Figure 4.3 indicates the condition that should exist at this point.

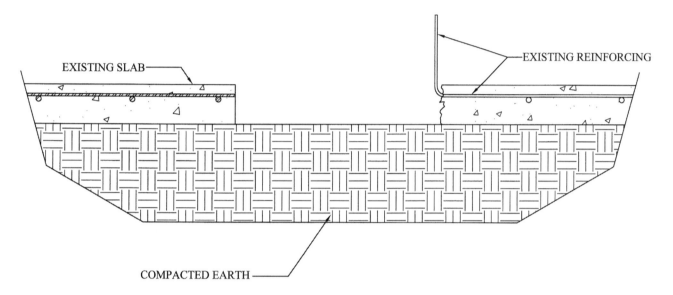

Figure 4.3 Slab preparation should look like this.

Excavate for your new concrete, making certain to remove at least 4" of material underneath the existing slab for bearing, then install your expansion joint (a couple of pieces of expansive foam typically used for "sill seal" is perfect for this purpose), and bend the rebar down into place. Pour your new haunch, using the existing concrete surface as your screed. When it's cured, you can clean the edge and caulk.

If you find no rebar in your slab, then you can just saw-cut the concrete in the second location full depth and remove the concrete in larger sections. In this case, when you excavate, you should make certain to underpin the existing slab a minimum of 8". Figure 4.4 indicates a section through the finished product in this condition.

Prior to pouring your concrete haunch, you will want to cover the edge of the existing concrete with a bonding agent made for that purpose. Sika Corporation manufacturers a product called *SikaDur Hi Mod*, which is an excellent bonding agent that I have had great success with in the past. Simply follow the manufacturer's instructions for a fail-safe installation.

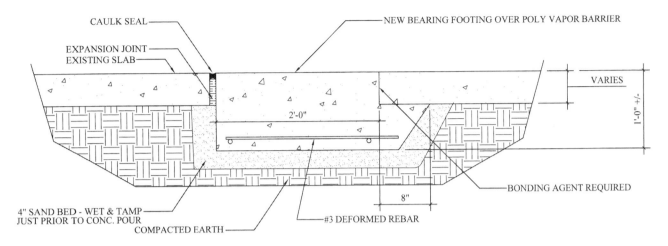

Figure 4.4 An example of slab preparation, minus reinforcement.

Floating Concrete Slabs

Floating concrete slabs are also excellent products for sound isolation, if designed and constructed properly. They are also very expensive, a tremendous amount of work, and on the low end of recommendations I would make for a home studio. I will include information here, but unless you are very serious about spending the time, effort, and money to obtain nothing shy of professional results in your home studio, I would not suggest that you bother including them for consideration. The slabs must be designed for a center frequency of no greater than 10Hz, and they only really make sense if you are going to achieve the same results with your wall and ceiling assemblies as you do with your slab. For this purpose, picture seven layers of drywall on your wall and ceiling assemblies.

Figure 4.5 is one product manufactured by Mason Industries for the purpose of constructing an elevated concrete slab.

Figure 4.5
To construct an elevated concrete slab, you can use the Mason jack-up floor slab system.

This is a lift slab product, utilizing a neoprene puck as the spring in the system. With this product, you can pour a new slab directly over an existing slab (thus saving on the bottom formwork required to pour an elevated slab in place). Figure 4.6 is another member of the Mason Industries family. It's an FS Spring Jack, which utilizes a coil spring instead of the neoprene isolator.

Both systems will work very well when the slab is properly designed, but for the purposes of examples in this book, we will use the neoprene puck system.

Figure 4.6
An example of spring jack floor mounts.

There's something you need to understand if you plan on using this method of construction. Lightweight concrete weighs as much as 115pcf (pounds per cubic foot). The spacing of the supports for this particular system can be as great as 54" on center. That translates to 20.25s.f. (square feet) per support, which, with a 6"-thick slab, would be 1,164 pounds per support. That's just for the slab. In addition to this, you will have the load imposed by the new walls and ceiling. It's quite possible that a simple slab would not be able to take this loading; for example, that you could have cracking and settlement of the existing slab, which would destroy your new work. We typically pour reinforced concrete with a series of mini haunches to carry pin loads like this in new construction. It is very important that you contact a structural engineer to assess your existing structure if you plan to use this method as a part of your studio design.

I promised you I would not bury you in math in this book, and I intend to keep that promise. There are calculations that need to be performed to design your slab to the required 10Hz center frequency, but I'm not giving them to you. It is so much easier for you to contact the manufacturer directly, and they will happily perform those calculations as a part of their service in selling their product to you. You will be required to provide them with detailed information about your construction, so it isn't as if you can decide today that you want this without planning your entire studio at the same time.

The design of these slabs is based on the slab weight itself, the weight of all walls and ceiling loads (including finishes), and the weight of equipment to be placed permanently within the room. (This is what makes up your actual and superimposed dead loads.) Then you have to figure out what your live loads will be. This should be a fairly accurate assessment of people and gear that will be brought into the space.

Floating slab isolation is based on a Mass Spring Mass (MSM) system. The first Mass is the element the slab will be supported by. The Spring is (in this case) the Dupont Neoprene pad sitting below the lift mechanism. The second Mass is the elevated slab itself. The reason for identifying all of the imposed loads properly is due to the fact that for an MSM system to operate properly, you must load the spring just enough to fully engage it. Overloading (overcompressing) the spring will cause the system to fail because the sound will transmit directly through the spring. Underloading (thus undercompressing) will cause the same effect. It's the combination of the mass and the air gap that provides your insulation. You do not want the spring to create a flanking path of its own.

The air gap is an important factor to take into consideration when using this system. For example, tests performed at the Riverbank Laboratories in Geneva, Illinois, proved that pouring a 4" slab over an original 6" slab only increased isolation from STC 54 to 57. Adding a 2" sealed airspace raises the STC to 79. Doubling the airspace to 4" increased this to STC 82.

A Comment on the Value of Air Spaces

Theoretically, a doubling of air space would add 5 STC to an assembly, but other things also affect the outcome. In this case, resonance takes over, so the effect is closer to 3.

In reality, if you have a slab with a resonant frequency of 10Hz and an STC rating of 79, you will be stretching it to achieve this with the remainder of your room construction.

After your design is finalized, you'll lay out your grid, install your edge forming, and place the lifts on the intersecting points (see Figure 4.7). The spacing at the slab edge will often be closer than the spacing in the field (center) of the slab. This is due to the load that will be imposed on the slab edge by the walls and ceiling.

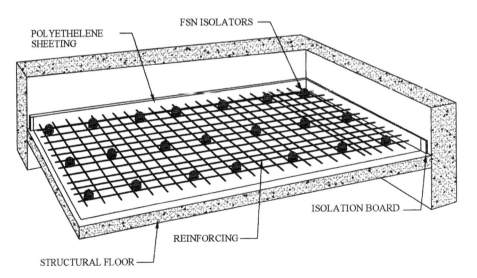

Figure 4.7
Placement of isolation materials should look like this illustration.

Rebar is then installed and supported by the hooks you see on the lifts. You need to make certain to tie your rebar together using "tie wire," which can be purchased in rolls at any lumberyard. A couple of turns around the rebar, where two bars intersect, and a few twists of the wire together to lock it into place, and you're all set. This will ensure that the rebar stays put when you place the concrete. It won't do you any good if it's lying on the bottom of the slab when the pour is complete.

Edge forming isn't all that difficult in this case. You can do this with 2x6 or 2x8 framing lumber for the sides (depending on your slab thickness) with 2x4 framing to lock them into place. The tops should be straight and true; then they can be tied in with 2x4 braces. You can use a builder's dumpy level or a laser (both are available from any tool rental company) to strike a level line completely around the slab. Mark the sides of the forms and then use 6- or 8-penny finish nails to mark the finish height of your slab. These are easier to pour than trying to use a line that gets covered with concrete and becomes invisible. Just tap them about 3/4" into the form to lock them securely into place.

A lot of people say that floor finish depends on floor covering, or that if you plan on covering it up you don't have to be too fussy. They're wrong. The better the quality of finish you provide to the floor, the flatter the floor will be in the end. And flat is important if you want to put down wood flooring and not have it make funny noises. I'm not going to try to teach you here how to pour and finish a concrete slab. If you don't know that already, you would be much better off paying someone experienced to do the work for you. Figure 4.8 shows the conditions expected during the concrete pour. Make sure to install the rubber plugs prior to the pour, which ensures that no concrete can enter the threaded chamber of the slab lift.

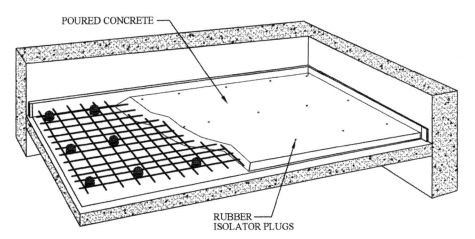

Figure 4.8
This illustrates a concrete pour in progress.

One trick you can use to help you along is to rent a laser level. You can set this to your finish slab height and use it to maintain the proper screed height as you go along. We don't generally use pipe screeds anymore in the industry; this equipment is the only way to go. If you do this yourself, make sure to set the instrument to a fine setting ($1/8"$). This will help ensure a flat surface when you're finished.

You are going to have to let your concrete fully cure before you begin to lift the slab. Don't even consider lifting the slab for the first 28 days. Twenty-eight days is typically the point where concrete will cure to its designed strength. In order to ensure that the concrete cures properly, you will want to flood it with water after finishing and then cover it with a layer of poly. Be certain to check it at least once a week (or sooner) to make sure it stays wet; this is an important part of the curing process. There are also curing compounds you can use for this purpose, but the least expensive manner is simply water.

After the concrete finishes curing, remove the forms, lift the slab, and you're ready to begin work on the studio walls. Figure 4.9 shows a slab in the elevated position before any additional work is in place.

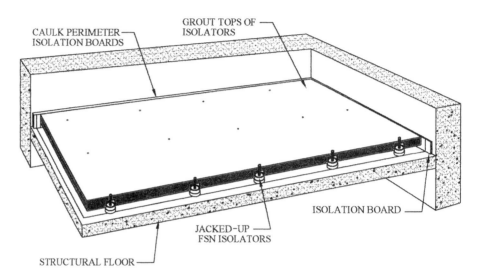

Figure 4.9
Here is an isolated slab in its elevated position.

For the actual results you might expect with this system, we provide the following data from tests performed for Mason Industries.

Description of Test Specimen

The test specimen consisted of a 4-inch thick reinforced concrete slab, with an average weight of 50lb. per square foot, supported on a grid of Mason Industries Type FSN-1336 and 1337 mountings with Type EAFM neoprene elements. The mountings are spaced on 24-inch centers, supporting the 4-inch thick "floating slab," 2 inches above a 5-1/2lb. per square foot elevated composite concrete slab. The floating slab was finished in the studio with a layer of 1/8-inch linoleum flooring. The floating slab was isolated along its perimeter with Type 34AFG-10, 3/4-inch thick, 10lb. density fiberglass boards. The test surface area was 1,054 square feet.

Test Methods

A) Impact Test

The test method used in obtaining the data is in accordance with ISO (International Standards Organization) Recommendation R-140-1960, "Field and Laboratory Measurement of Airborne and Impact Sound Transmission."

The data obtained were corrected to a reference room-absorption of 10 square meters. In accordance with the R-140-1960 Standard Recommendations, the absorption of the receiving room was measured by recording the acoustical decay rate.

Test equipment consisted of the Bruel & Kjaer Tapping Machine Type 3204, which was placed at two (2) positions on the test floor. Sound pressure levels were measured in one-third octave bands in the receiving room by using the General Radio Noise and Vibration Analyzer Type 156A, and corrected for ambient noise levels in the receiving room. Test results represent the arithmetic average of the two positions measured. These positions are indicated with "X" marks on drawings Z-1075-1.

B) Airborne Sound Transmission Loss Test

The test was conducted in full conformity with Section 6 of American Society for Testing and Materials Designations E 336-71, Standard Recommended Practice for Measurement of Airborne Sound Insulation in Buildings.

Acoustical test signals of prerecorded 1/3 octave band random noise were generated in the source room with loudspeakers placed in such manners as to generate a diffused sound field. Four loudspeakers were used consisting of two 8-inch diameter acoustically suspended speakers, and two horn speakers.

Test frequencies were 1/3-octave band frequencies between 100 and 5,000Hz inclusive. Sound pressure levels were measured in the receiving room in 1/3 octaves by using the General Radio Noise and Vibration Analyzer Type 1564A, and corrected for ambient noise levels in the receiving room. The data obtained were corrected to a reference room-absorption of 10 square meters. In accordance with the ASTM E336-71 Test Standard, the absorption of the receiving room was measured by recording the acoustical decay rate.

Test Configuration

The test specimen separates the 25th floor TV studio and elevator equipment room on the 24th floor. For the impact test, the receiving room was the elevator equipment room. Measurements were made in the elevator equipment room in the late evening hours with all elevator equipment, as well as ventilation equipment, shut down in order to permit measurable sound pressure levels tied to the Tapping Machine.

For the Airborne Sound Transmission Test, the elevator equipment room was used as the source room and the TV Studio was used as the receiving room so as to minimize the amount of time elevator equipment was shut down. Partitions around the TV Studio were partially supported on the test specimen floor as shown on Mason Industries drawing Z-1076. Furthermore, the ceiling over the TV Studio consisted of a resiliently suspended gypboard ceiling with a layer of 4-inch thick glass fiber blanket laid over the top.

Test Results

A) Impact Noise Rating (INR)

Sound pressure levels at 1/3 octave intervals, normalized to 10 square meters, are as follows:

Center Frequency (Hz) in One-Third Octave Bands	Sound Pressure Levels (dB) Normalized to Ao = 10 M sq
100	51
125	49
160	48
200	48
250	47
315	42
400	36
500	34
630	28
800	29
1000	28
1250	24
1600	24
000 2500 3150	*
Impact Noise Rating	INR + 24

Note: * Denotes sound pressure levels due to Tapping Machine being below ambient noise levels for those frequencies indicated.

B) Field Sound Transmission Class

Sound transmission loss values are tabulated below at the 18 standard test frequencies.

Center Frequency (Hz) in One-Third Octave Bands	Transmission Loss (dB) Normalized to Ao = 10 M sq
100	47
125	48
160	50
200	54
250	60
315	66
400	71
500	79
630	85
800	90
1000	95
1250	92
1600	92
2000 2500 3150 4000 5000	*
Field Sound Transmission Class	FSTC – 71

Note: *Denotes transmitted sound pressure levels being below the ambient noise levels for those frequencies indicated."[1]

One of the things I like about this test data is that it gives us a chance to look at the differences between laboratory tests and real-world field conditions. The expected Impact Noise Rating (INR), based on lab testing, was INR + 17, which in the field test yielded a result of INR + 24, a fairly significant increase in isolation. This was attributed to 1/8" linoleum flooring placed on the field specimen versus a bare concrete slab in the lab. Also, part of the increase was apparently due to the supporting slab itself, which was a rigid structural slab (in the real world) compared to the precast "T" section concrete panels used in the laboratory tests. The impact rating could obviously go in either direction, based on field versus lab structural differences.

The difference in STC rating was as follows:

Laboratory STC rating was 79 vs. FSTC rating of 71.

(The F in FSTC stands for "Field," i.e., the real world.)

Once again, this is a very significant difference. An 8dB reduction (in isolation) of the final assembly in the field application (although not surprising to me) means that the noise making it through to the other side is more than two times louder than expected by lab results. The apparent reasons for this were that the walls surrounding the studio were capable of less sound attenuation than the slab itself. The partitions in the studio were drywall vs. masonry partitions used in the lab. Also, a single layer of gypsum was installed in the ceiling of the studio. This was not capable of matching the double layer used in the lab. The walls in the studio were constructed to the underside of the structure above. The single-layer drywall ceiling (coupled with the wall assembly's attachment to structure) allowed sound to travel through the building structure outside of the TV studio and then into it. This is an example of flanking noise affecting your total sound isolation.

The third item has to do with something I've mentioned earlier in discussions here and that involves the quality of work performed in the field by the professionals in the construction industry. In lab tests performed by a bona-fide laboratory, the construction is as close to perfect as is physically possible. Care is taken to provide properly installed and sealed assemblies. In the real world (the world that your construction will take place), contractors have a vested interest in getting in and out of the project. It's not that they don't really care about the finished product, but they don't necessarily devote the attention to detail that takes place in that laboratory. Many times the person who bids the work never even sees the job. The workers in the field never receive the information they need to understand exactly what steps need to be taken or the order they need to be taken in.

In addition, they see some of what we (the owners and engineers) place importance on (which is that same care taken in the labs) as being less than necessary for a "decent job." They do not understand the importance, they think it's overkill, and they think that the small difference in quality won't mean a whole lot in the end. They could not be more wrong.

One other point I will make here before moving on is that the test result you see only relates to a floor-to-floor separation and not room-to-room. I can guarantee that the wall and ceiling construction used for this studio would not come anywhere near the STC 71 reported here floor to floor, and thus the money for the floating slab would have been wasted if that were a concern.

In the case of this TV studio, apparently their biggest challenge was noise being transmitted from the elevator machine room into the studio. In this case, it paid off. In your case, however, it's going to be a concern for you with your neighbors and family. So unless you plan on going the full route, meaning that unless you plan on building a "bomb shelter" to record in, don't waste your money by taking this path. You can get some fabulous results with a floating room on a floating slab if you need them, but you have to make it all work as a whole—one piece won't get it done by itself.

Floating Wood Decks

I will spend just a moment discussing decks because I hope to save you some money and aggravation. I wish I had a dime for every time I've seen someone ask (on an Internet site) how to calculate the spacing of pucks for a floating wood deck. I know immediately that they do not have a clue what they're building. Someone, somewhere, just told them that they needed to float a deck in order to get real isolation, and they listened.

Reality is that you can't get isolation for a recording studio with a simple elevated wood deck. Period.

Elevated wooden decks are great for impact noise, so they'll work well in dance studios, bowling alleys, condos, etc., but they do not have enough mass to isolate a recording studio properly. The frequencies of the bass drum and bass guitar are too low to be handled with a deck of this nature.

Comment from Mason Industries Regarding Wood Framed Floating Floors

This is another bit of information from the Mason Industry site:

"It is often necessary to provide a wooden floating floor rather than the heavier concrete construction with wood topping. Cost or weight restrictions may be the factor. In older buildings, it is often necessary to improve on existing floors with a lightweight impact noise resistant construction. A resiliently supported wooden floor will reduce the rumbling noise of a bowling ball, the click, click of a woman's heels and that portion of a typical noise generated by a piano that travels down the piano legs and into the structure. It will offer only minor reduction of airborne sound, as there is insufficient mass in the surface. In some applications on stages or in rehearsal rooms, the primary purpose is relief and comfort for the dancers. Landing on concrete or hard mounted wood surfaces is very damaging to a dancer's feet and legs."[2]

Another problem with isolated wood decks (and this again relates to the lack of mass) is that the resonant frequency of the deck itself may create problems. As I pointed out with the elevated concrete decks earlier, the resonant frequency of the deck needs to be around 10Hz. With concrete, achieving that frequency is a matter of proper design; with a wood-framed deck, it's a matter of being so close to impossible that it isn't worth the effort or expense.

For example, I recently came in contact with someone who owns a professional studio, with some of their rooms rented for space for practice, and others for tracking and control rooms. They were inquiring about the density of the insulation they should use in these already-built isolating decks they constructed. I explained to them the problems they were going to experience, but they apparently did not believe me. They completed their construction (based on "expert advice" they received from others) and then came back to let me know about the problems they were having with these "drumheads" they built. Frequencies between 200 and 400Hz were being amplified by the deck construction, and they wanted to know what they could do to solve the problem. My answer to them was to either remove the decking and fill the frame completely with dry sand or remove the decks in their entirety.

Seeing as this is a (roughly) 3,000s.f. facility, that translates to a lot of wasted materials and labor any way you look at it. All of the decks were filled with mineral wool and then multiple layers of decking plus finished floors. Picture the cost of that construction and now add to it the cost of removing those finished floors and decks—pull out that mineral wool (throw it away because it isn't going to do you any good anymore)—now bring sand in and begin finishing these spaces all over again.

So please, I don't care what your friends tell you. I don't care what some "expert" on the Internet tells you. Save your money. It's too precious to just throw away.

Sand-Filled Wooden Decks

Unlike elevated wood decks, sand-filled decks can be of some benefit. One benefit is the creation of a pathway below the floor to route low-voltage (LV) wiring (through the use of PVC conduits). It's a big advantage if the LV wiring can be separated from line-voltage cable runs (plugs, lighting, etc.).

You can also gain some small benefit from an isolation point of view, which is due to the ability of the sand to damp sound vibrations.

However, this is a very heavy assembly (sand alone weighs roughly 100pcf) and (as in the case of concrete slabs) will require a structural engineer to confirm that you can safely install this on your existing structure.

Wall Construction

Wall construction is not really all that involved, provided you pay attention to details. Your biggest challenge is to determine what level of isolation you need, the amount of mass you need to handle that, and then to make certain to put everything in the right location.

The right location. Think about that for a moment. There is actually a right location for materials.

Yup, there sure is.

Look at Figure 4.10 for a moment.

Note that the wall assembly third from the left has an STC rating of 40dB. It is constructed using two separate wall frames with four layers of drywall total and with insulation installed in the cavities of both walls. The drywall is installed on both faces of each wall assembly.

If we remove one inside layer, note how the STC rating rises from 40 to 50. Remove the second layer, and it rises another 7dB of isolation to 57. Now take those two layers and install them over the two faces of the outside wall, and you add another 6dB of isolation for a total of STC 63.

Chapter 4 Floor, Wall, and Ceiling Construction Details

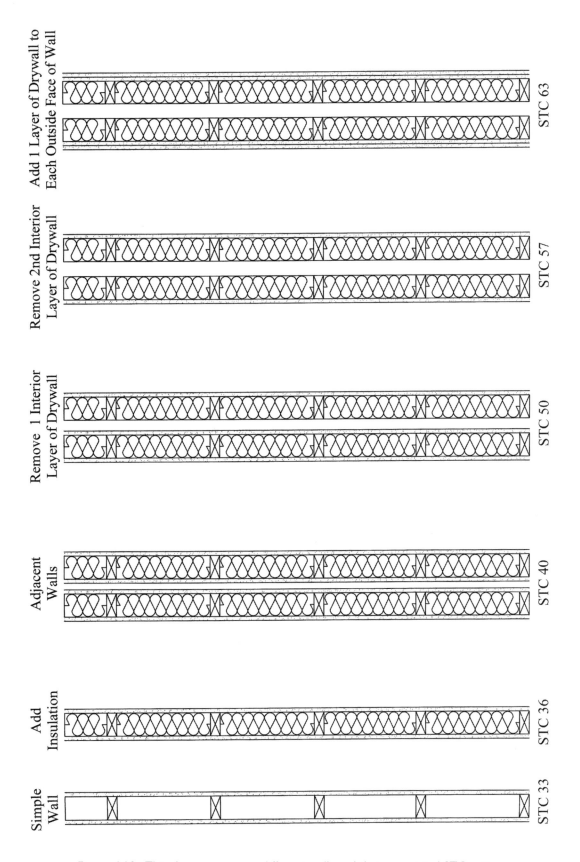

Figure 4.10 This shows you some different walls and their associated STC ratings.

63

This is a 23dB increase in isolation using exactly the same materials in different locations!

That's an amazing increase in isolation, and it didn't cost a single additional penny to gain it.

This appears to fly directly in the face of what I told you earlier—that each doubling of mass adds roughly 6dB of isolation to your wall. However, the reason for this has nothing to do with the amount of mass and everything to do with the location of that mass.

Consider that each different location of the drywall constitutes a leaf. In the case of the STC 40 wall assembly, you have two walls with one layer on each wall face, thus, a four-leaf system. You count the leafs thus: Mass, Air, Mass, Air, Mass, Air, Mass. Each area of mass (drywall) equals one leaf. When you remove one inner face, it becomes a three-leaf system and then a two-leaf system when you remove the last inside face of drywall. A two-leaf system with $1/2$ the mass of the four-leaf system will provide almost eight times greater isolation. It is at that point that the doubling of the mass becomes what I told you earlier. Note the 6dB increase from the STC 57 to 63 when the mass is doubled and put in the correct location.

So get a picture firmly in your mind—two-leaf wall systems are good, two-leaf wall systems are your friends. Three- and four-leaf wall systems are evil devious fiends that suck up the money in your wallets and give you less than nothing in return.

So when you build, strive for two-leaf wall systems and nothing else. These wall assemblies are Mass Air Mass (MAM) Systems. With MAM systems, the air between the leaf walls becomes the spring. Thus, sound pressure builds up within the space striking the wall surface and causing it to deflect inward. This compresses the air within the space, which causes the outer surface to deflect outward. The compressed air becomes the spring. The transfer of sound is its vibrational energy traveling through the spring from one layer to the other. If this energy rested within the frequency range of the wall assembly, it would cause a fairly significant lowering of the TL (transmission loss) of the assembly. The more mass you add, coupled with the air space and the stiffness of the assembly, all help to lower the resonant frequency of the assembly.

Existing Walls

Existing walls are going to be your biggest challenge. The exterior walls are probably a long way away from being airtight. Interior walls might be running parallel with a corridor, and perhaps that corridor is already only 36" wide, which means you can't add more mass to the corridor wall without creating a violation of the building code by creating a corridor whose width is too narrow. Maybe your building is in a basement and has to contend with windows, perhaps large gaps between the foundation and the plate that carries the deck framing. Or you are using an existing bedroom that the building official requires to remain as is, with the egress windows intact.

In all these cases, there are solutions to your problems. In the case of the concrete foundation, you can fill those cracks between the plate and concrete with mineral wool. Just pack it well to make certain you have good density and then caulk the joint between the two with acoustic caulk.

Figure 4.11 indicates both the original condition of the foundation and the finished seal.

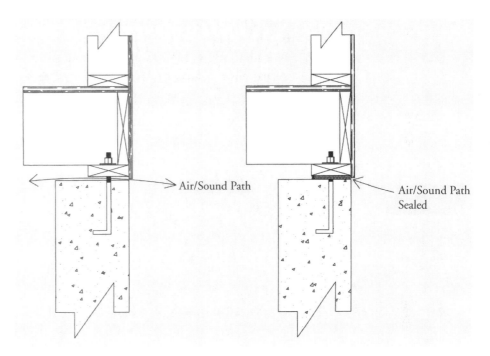

Figure 4.11
Sound leak at existing slab—before and after seal.

In the case of that corridor wall, you can remove the drywall on your room side and install additional mass within the wall cavity itself, which will go a long way toward helping your isolation.

We will look at the actual construction of these assemblies in the next chapter. For now, let's look at the various types of wall assemblies and their possible benefits or drawbacks.

Wood Walls

Wood is one of my personal favorites when it comes to studio construction. There's just something beautiful about a properly framed wood wall. In the days when I used to swing a hammer, we had a saying about wood framing, "It should be so pretty you don't have to cover it up."

Let's focus on the options.

Single-Wall Construction

Single-wall construction is what you're used to seeing in your home. It's a simple wall—typically a single 2x4 or 2x6 plate—with studs ranging from 16" to 24" on center (oc). It probably has double top plates and usually a layer of 1/2" gypsum drywall mounted on each of the faces.

You already know it's not a good isolator, or you wouldn't be reading this book.

Resilient Channel/RISC Assemblies

Take that same wall and add resilient channel to one side of it (after removing the existing layer of drywall). This decouples the layer of gypsum on that face from the framing behind it. With that same layer of 1/2" gypsum, you just increased your isolation from somewhere between 3 and 5dB.

There are many different types of resilient channels on the market today, as well as materials you might mistake for resilient channels, for example, hat channels or Z channels. RC-1 is a single-leg channel primarily used for walls. RC-2 is a double-legged channel primarily used for ceilings.

Figure 4.12 is a picture of various types of resilient channels. I have also included a section of hat channel. Note that hat channel does not have any holes or slots along its "leg." The slots and holes are one of the main reasons that the members isolate in the manner they do. Without them, isolation is nil.

You can use either RC-1 or RC-2 for walls and ceilings; however, RC-1 used on a ceiling will not support the same load as the RC-2.

Figure 4.13 shows RC-1 installed properly on 2x4 framing. Note that the flange is installed in the bottom position. It is important that these products be installed correctly to ensure that they isolate properly.

Make certain to be very careful that you do not use the hat channel or standard Z channel instead of RC, because the net effect will be no gain in isolation with a lot of added material and labor costs. And, amazingly, there are a lot of drywall contractors out there who can't tell the difference between the two. Whatever you decide to purchase, make sure to ask for (and receive) the technical data as a part of the purchase. This will ensure that you're getting the right material.

RISC 1 clips (resilient sound isolation clips) are another excellent way to decouple drywall from a structure (see Figure 4.14). These products use a rubber isolator to decouple the clip from the stud, and a hat channel is then mounted into the clip to attach the drywall. This unit will typically provide around a 6 to 8dB increase in isolation over what you can expect using standard resilient channel. Once again, there are several different manufacturers of this type of product.

Figure 4.12
Resilient channels.

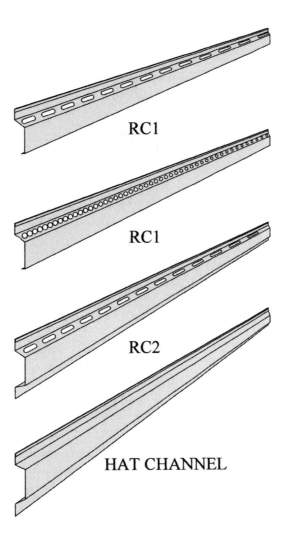

Figure 4.13
Resilient channel installation method.

Figure 4.14
Resilient sound isolation clips (RISC-1).

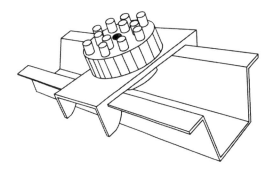

Dual Frame Assemblies

Dual wood-framed wall assemblies come in a few different types. One of these is a staggered stud configuration, with single top and bottom plates (typically 2x6 or 2x8), with the studs staggered from one another on opposite sides of the wall. With framing 16" oc, that would place the studs on 8" centers and 12" centers for 24" framing. Figure 4.15 is a view of a typical staggered stud system.

The advantage is greater isolation than the use of single-frame walls, decoupling the drywall surfaces from one another reduces the need for RC or RISC systems.

The disadvantage is that the common top and bottom plates become the weak points of this system, effectively acting as a bridge from one side to the other. This direct passage weakens an otherwise good system.

True isolated wall assemblies use separate top and bottom plates, as well as separate studs, which effectively completely decouples each wall from one another. Figure 4.16 is an example of double-framed wall assemblies.

Chapter 4   Floor, Wall, and Ceiling Construction Details

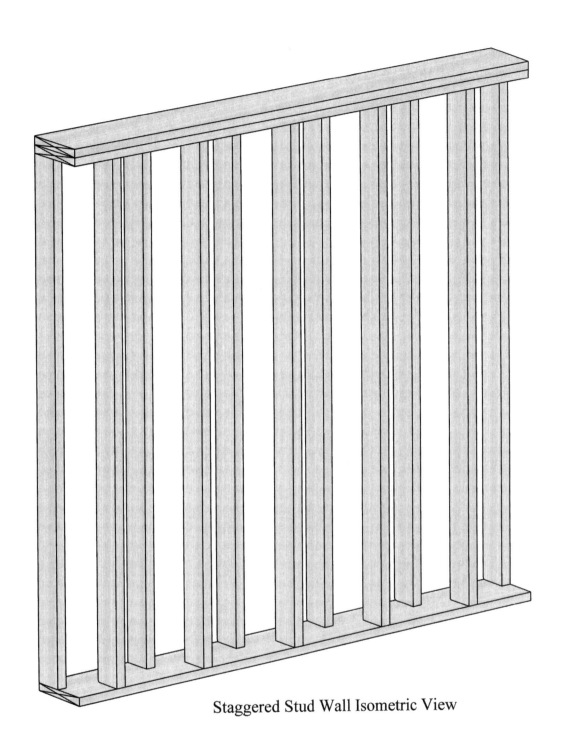

Staggered Stud Wall Isometric View

Staggered Stud Wall Plan View

Figure 4.15   Staggered stud framing.

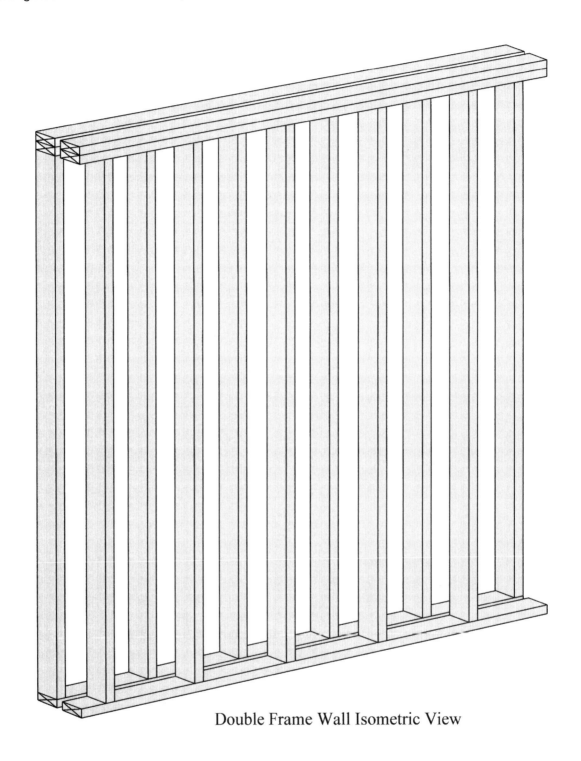

Double Frame Wall Isometric View

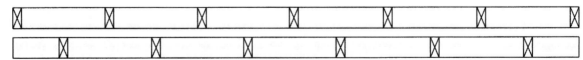

Double Frame Wall Plan View

Figure 4.16 Double wall framing details.

One point I want to be sure you get is that there is no advantage to using isolating clips or resilient channel with a true double wall assembly. Don't waste your money by thinking that more is better in this case. The only thing that should be more with a double wall system is mass and (perhaps) constrained layer damping.

The advantage to this is that it's not much more money than a staggered stud wall assembly, and it can achieve much greater isolation that any of the clip systems. The disadvantage is that it takes up more real estate (by this I mean room space) than the RISC, RC systems, or a simple staggered stud system with common top and bottom plates.

However, when isolation requirements are high, and you weigh the isolation benefit vs. the cost per square foot of construction, then the added isolation can be such a dramatic increase that the added cost is well worth the investment. You will reach a limit with the other systems (based on common connection points between the adjacent spaces), such that the flanking action of the frames will become the weak point. This is one of the great strengths of a double frame assembly.

Now, let me clarify that last statement. In all assemblies—unless care is taken to address flanking paths—there will come a point where the passage of flanking noise will take place over the value of the sum of the separate assemblies. In actuality, flanking will always be the deciding factor in the total isolating value of an assembly. But you can go a long way toward raising the level of isolation by making certain you do not have any hard connections between adjacent structures. As an example, it would do you no good to build wall and ceiling assemblies with 70dB isolating values if you're on the second floor of a building and are sitting on a deck assembly, which is structurally connected to other parts of the building.

The sound that travels through the building structure will not decrease just because you continue to add more mass to the walls. If you can't create a condition where all of the elements of the assembly are balanced, then you will soon reach the point where you are throwing money at a problem without moving any closer to solving that problem.

Steel Framing

Steel framing has some advantages over wood framing. It doesn't shrink, twist, or burn. It's a very stable material with uniform dimensions. Perfect fits are not as critical with steel framing as with wood. Fastening is as simple as screwing in a couple of self-tapping framing screws.

When we refer to heavy-gauge steel framing, it's any material with a gauge greater than 25, whereas 25 gauge or lighter framing is considered lightweight steel framing.

Light-gauge steel framing has an advantage over wood framing or heavy-gauge steel framing because a wall constructed with it has a greater STC rating than a similar wood or heavy-gauge frame. So a simple wall framed with light-gauge framing and drywall applied directly to both sides will achieve the same STC rating as a wood frame wall with resilient channel installed.

The disadvantage of light-gauge framing is that it will not carry loads that wood framing will carry with ease. Heavy-gauge metal framing will carry the required loads, but it is generally more expensive than similar walls manufactured with wood and (from an isolation point of view) will provide the same results as wood.

A good read if you want to use steel frame construction is the *USG Gypsum Construction Handbook*. This handbook is a great source of information on this type of construction. It can be located here:

www.usg.com/resource-center/gypsum-construction-handbook.html

Another excellent reference for steel frame construction is the *Dietrich Trade Ready Design Guide*, which is full of engineering information as well as details for construction related to steel framing. It can be found at:

www.dietrichmetalframing.com/library/trdesignguide.asp

Masonry Construction

The use of concrete and masonry block to construct isolation walls is another effective means of accomplishing your goals.

In some parts of the world, masonry construction is preferred over wood due to the difficulty in obtaining structural wood framing materials and because it is the least expensive method of construction.

Standard hollow masonry blocks are readily available in nominal 8x8x16 inch dimensions. You'll take a closer look at this in Chapter 10. See Figure 4.17 below to view typical block shapes.

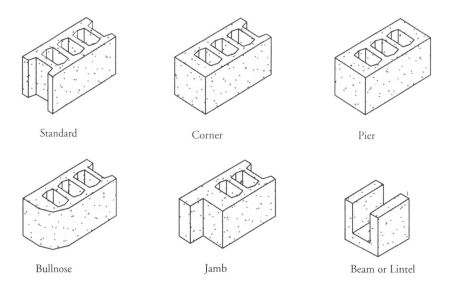

Figure 4.17
Standard concrete blocks.

Standard Corner Pier

Bullnose Jamb Beam or Lintel

Ceiling Construction

"One man's ceiling is another man's floor." Paul Simon, 1973.

Isn't it always the way? You want to make music, and some inconsiderate human being wants to quietly read a book, sleep, settle a baby down, etc. (Fill in the blank based on your situation at home.)

Constructing an isolated system is only as good as the weakest link in the assembly—with the greatest challenges being spaces directly above or below you.

Let's take a look at those challenges and methods of maintaining isolation.

Working with Existing Ceilings/Floors

What steps can be taken to obtain isolation when building below other living spaces? First and foremost, keep your eyes on the basics—mass, mass, and more mass.

A typical deck in a modern home is generally ³/₄" T&G (Tongue and Groove) plywood installed over floor joists or Truss Joist (TJIs), with possibly pad and carpet or hardwood floors installed directly over this. Kitchens and bathrooms will generally have an additional layer of ³/₈" plywood installed over the ³/₄" deck, prior to the installation of resilient flooring or tile.

Figure 4.18 shows you some typical floor assemblies. Included are the approximate dead loads associated with these floors. Remember that the starting point for determining what may (or may not) be added to any structure begins with the existing structural capacity and weight.

One of the problems with just adding mass to the bottom of the existing joists is the transfer of sound directly through the joists themselves. Thus, the ceiling within your space has to be structurally decoupled from the floor above in some manner.

In addition to this, additional mass should be installed (if structurally possible) on the floor above. Remember that mass on both leaves of an isolation assembly is important. This floor/ceiling assembly is only different from your walls (acoustically speaking) in the sense that people walk on it, so the footfall noise (impact noise) adds another dimension to the challenges associated with isolation.

Once again, everything begins with a structural analysis of the existing assembly to determine what it is capable of carrying. Typical home design is mandated for a 10psf dead load and 40psf live load in kitchens, living rooms, family rooms, dining rooms, and game rooms. Bedrooms are typically designed for 10psf dead loads and 30psf live loads. The dead load is the weight of the floor joist, any decking above, and loads imparted by walls resting on the deck, along with the weight of floor coverings. The live loads are based on furniture within the space and the loads imparted by people within the room.

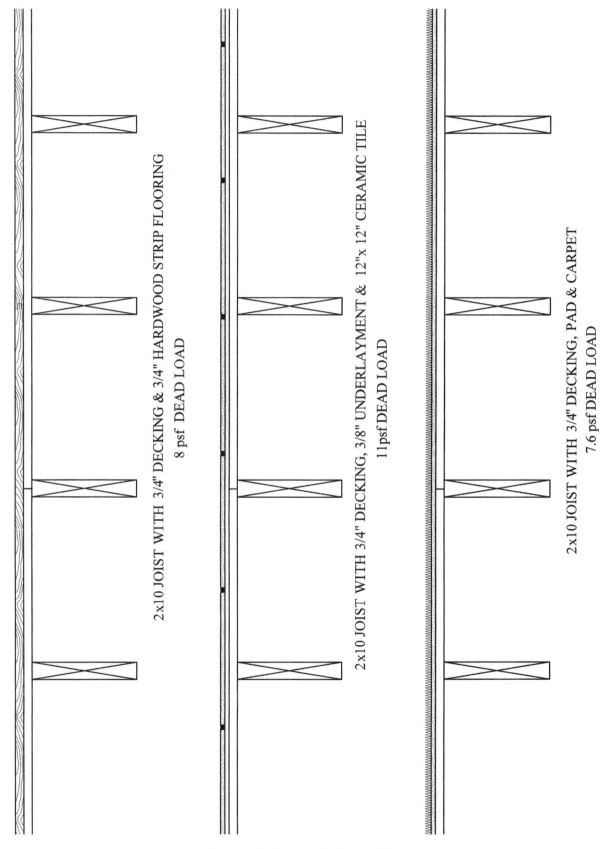

Figure 4.18 Varying deck assemblies.

Although you may think that 40psf is a lot of capacity, when you look into an empty room and take into account your furniture, your spouse, and possibly a few children, it can suddenly become a critical load when you have a family reunion and 35 or 40 people fill up that room. (By the way, a family reunion on my mother's side of the family may have as many as 350 direct family members show up, so big loadings are a real possibility.) The reason that I make this appear very important is because it really is.

Overloading a deck can cause long-term stress on it—with the result being a catastrophic structural failure when it receives a sudden additional load. I cannot stress the importance of this enough.

If you are designing a new home, it's easy enough to determine what you need and make provisions for this in your design. Perhaps the placement of your studio above the garage—with an isolating mudroom connecting the garage to the house—would make sense. Or you could make sure you had a foundation that was 9 or 10 feet in height to allow enough room for a decent ceiling height in a basement studio, while maintaining completely isolated construction of the two spaces.

In this case, you would also want to add mass to the floor above and possibly even a lightweight concrete self-leveling slab (also known as gypsum concrete), which helps tremendously with isolating traffic noise. We use this all the time in hotel construction with wood-framed deck assemblies, gaining additional isolation, as well as creating a fire barrier between levels. See Figure 4.19 for a typical view of a wood deck with gypsum concrete being applied.

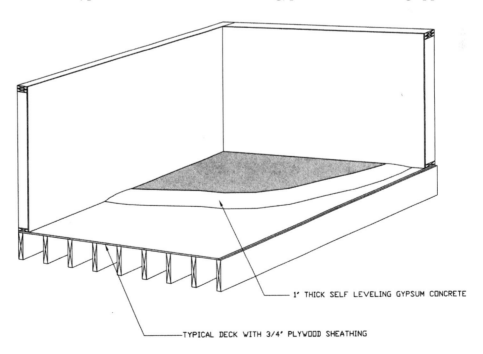

Figure 4.19
Gypsum concrete application over wood deck.

We'll look more closely at some isolating decks for new construction in the next chapter. However, in existing construction, we begin with establishing what we can safely carry in additional dead loads before proceeding.

Now, assuming that your engineer sees no problems adding load to the structure, you can begin by installing gypsum board directly against the bottom of your existing deck assembly. Fit the board so that it is about 1/4" from your existing floor joist, hold it in place temporarily with some 4d finish nails, and caulk all edges. Add as many additional layers as your joist can safely carry (see Figure 4.20).

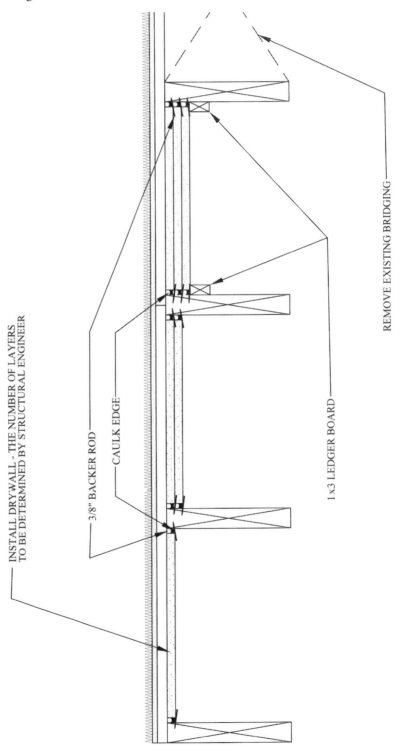

Figure 4.20
Added mass to existing deck.

Generally, this will require the removal of some cross-bracing members (known as bridging) in your deck assembly. Make certain to replace these, as they are critical (structurally) to your original floor design (see Figure 4.21).

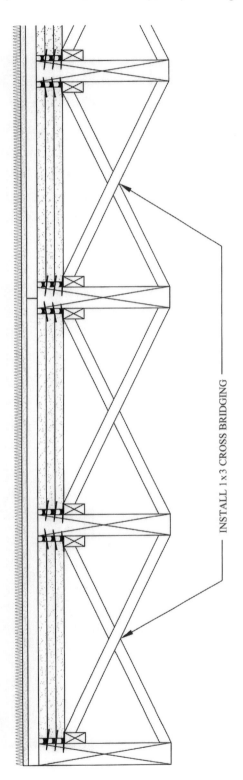

Figure 4.21
New bridging in place.

After all of this is in place (or back in place with the case of the bridging), you will want to insulate below that to the bottom of the floor joist. I have seen a lot of recommendations for this material to be rockwool; however, standard fiberglass insulation will do just as well and for less cost.

Below this, you are now ready to install your new ceiling. Let's look at some options.

Resilient Channel Ceilings

Resilient channel works as well on ceilings as it does on walls, although I would recommend the use of the double-leg system due its capacity to carry a greater load than the single-leg system. Refer to the manufacturer's recommendations for the maximum spacing and load for their particular product.

RISC-1 clips and hat sections can also be used on ceilings with the same results you get on walls. The advantage to using this system is that the drywall ceiling can run directly to the perimeter of the existing wall assembly. This solves issues with weak isolation points at that location, as well as creating a good seal for fireblocking. The disadvantage is lower isolation than totally isolated structural assemblies. Figure 4.22 is a typical installation using single-leg resilient channel on a ceiling.

Figure 4.23 is a double-leg resilient channel, and Figure 4.24 shows the conditions with RISC-1 clip assemblies.

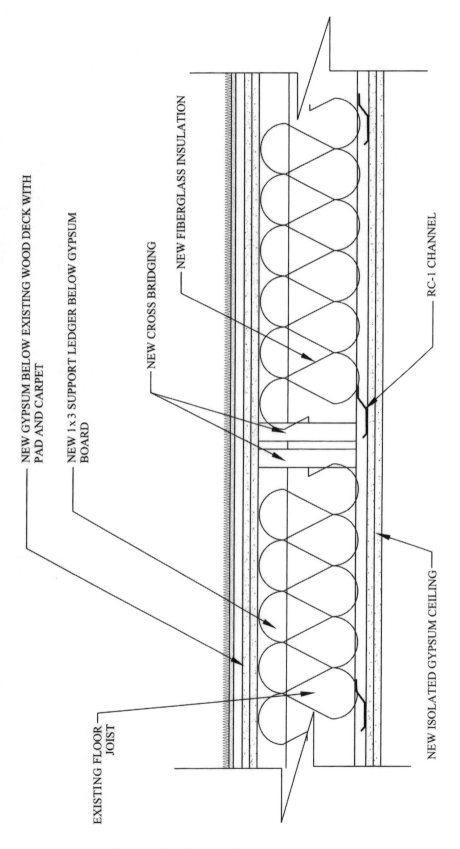

Figure 4.22 Ceiling with resilient channel.

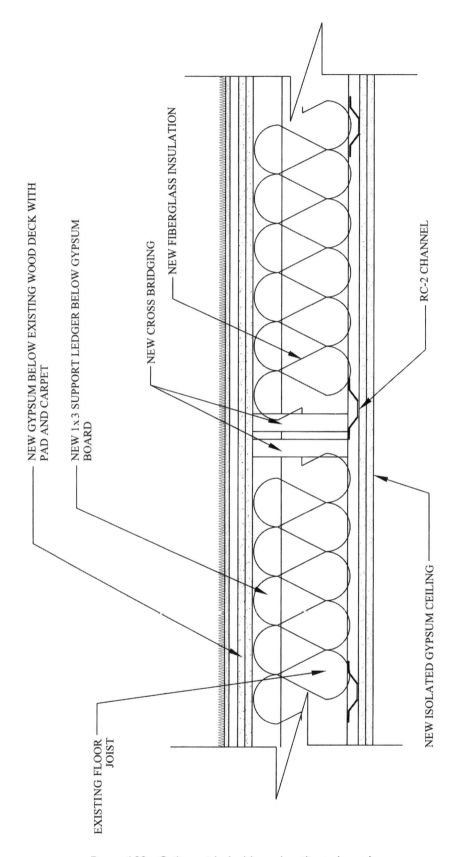

Figure 4.23 Ceiling with dual-legged resilient channel.

Chapter 4 Floor, Wall, and Ceiling Construction Details

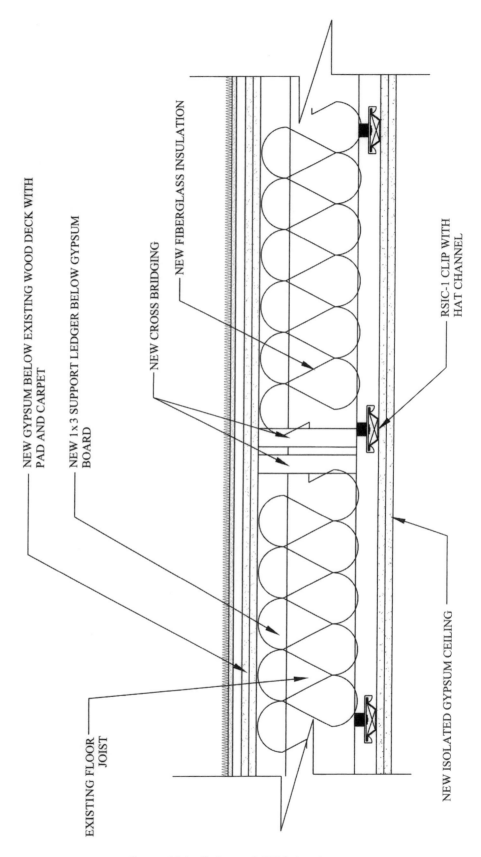

Figure 4.24 Ceiling with RISC-1 isolation.

Suspended Ceilings

Standard suspended ceilings, utilizing T-tracks with L-channels at the ceiling's edges and laying in tiles, are not of any use for the purpose of serious isolation and should not be considered.

Semi-Independent Frame Ceilings

A semi-independent frame ceiling would be supported by your interior walls at the edges—with interior supports added to transfer some of the ceiling load back to the existing structure above. This would generally be the case where you could not install a member large enough to support the entire load with a joist spanning the width of your room. This can be accomplished, with you still maintaining a good degree of isolation, through the use of isolation hangers carried by the structure above.

Once again, as with floating slab construction, it is very important that you take into consideration the weight being carried by the individual hangers in order to make certain you load them up enough to engage the spring action of the isolator. Too much or too little load, and they will transfer sound energy through to the structure above.

Figure 4.25 is a picture of a Mason Industries hanger made for this purpose. It's their WHR ceiling hanger.

Figure 4.25
WHR isolation hanger.

Figure 4.26 is a section through a deck (with the joist running lengthwise) that shows this hanger in use on a semi-independent ceiling assembly.

Figure 4.27 is a view of this with a section running through the new ceiling joist.

A downside to using this approach is (again) the load that it imposes on the existing structure, which will reduce the amount of mass you can add to the bottom of the deck above.

Chapter 4 Floor, Wall, and Ceiling Construction Details

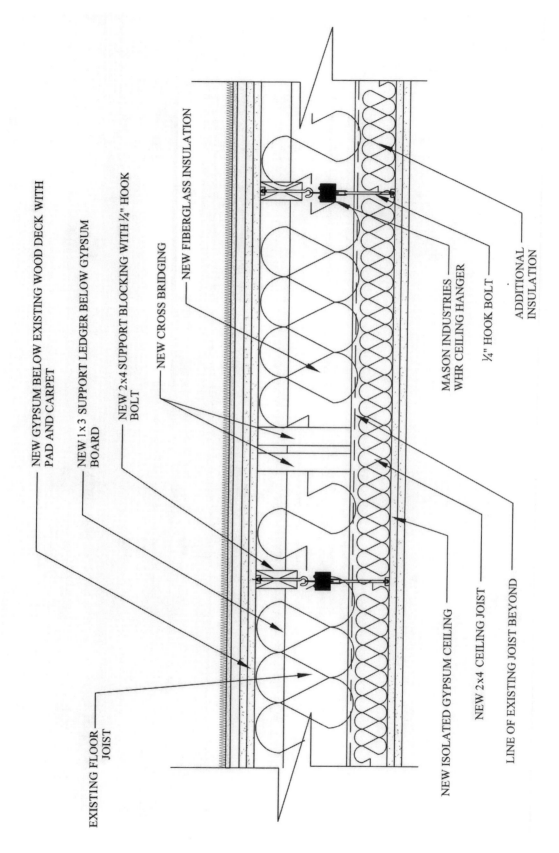

Figure 4.26 Semi-independent ceiling with isolation hangers.

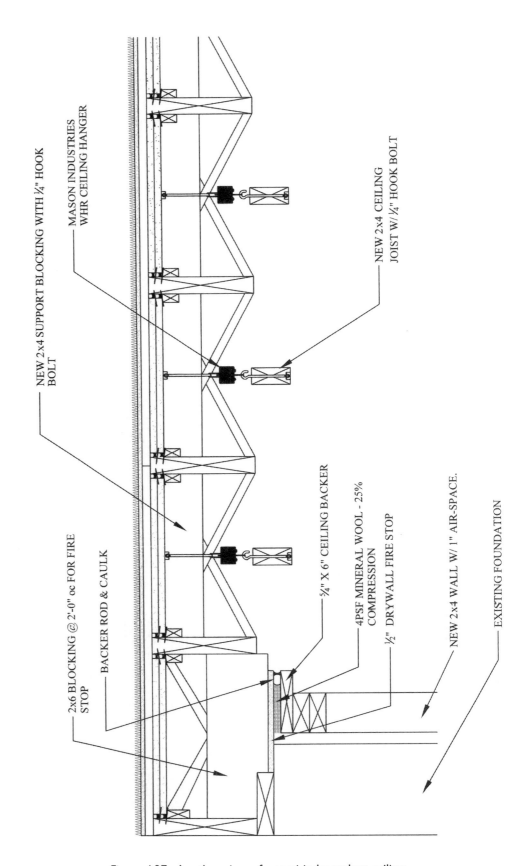

Figure 4.27 Another view of a semi-independent ceiling.

Independently Framed Ceilings

Independently framed ceiling assemblies carried by your new interior studio walls are the best isolation you can achieve. One of the big advantages of this system is the fact that it will impose no weight to the existing structure, so you can add that much more mass to that structure to help with isolation. Figure 4.28 is an independently framed ceiling below an existing structure. Note how the new ceiling joist is offset above the bottom of the existing floor joist. This is the reason I suggest you use cross-bridging rather than a solid block bridging for the existing assembly. Although solid bridging (a block made from the same dimensional material as the existing floor joist) creates a deck that is more solid than the cross-bridging, it will not allow the new joist to enter the pocket of the existing joist.

In the case of Figure 4.28 (with 2x10 existing joist and new 2x6 ceiling joist), the difference would be losing 3 $^{5}/_{8}$" of ceiling height versus 5" with solid bridging. In tight spaces, such as existing basements, that extra few inches could make a world of difference in a room.

Take a moment and focus on the firestop detail in Figure 4.28. It's important that you take care during construction to create assemblies that will not create chimney effects in case of fire. The $^{1}/_{2}$" drywall creates an effective firestop, yet maintains (as detailed) isolation between the two structures. Without this type of care taken during the construction process, a fire within the wall cavity could completely envelope the room in fire before you ever knew it was there. The creation of fire compartments within wall and ceiling cavities will provide a safe environment for you in case of an emergency.

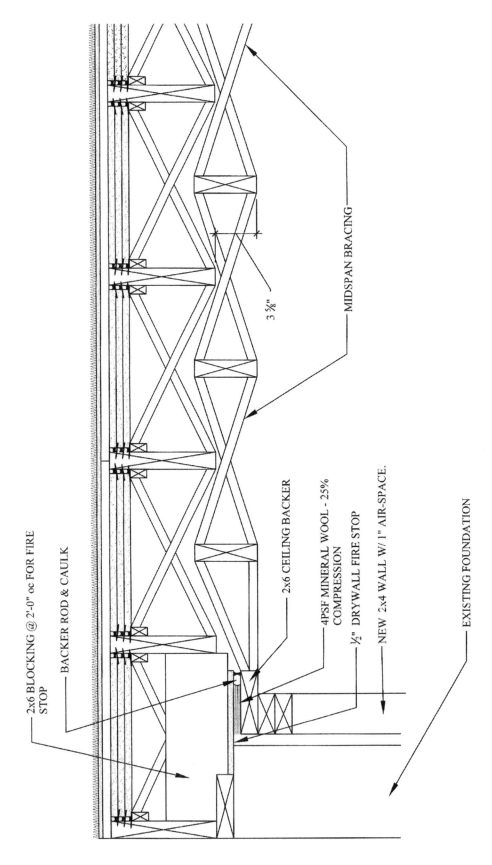

Figure 4.28 Independently framed ceiling.

Additional Isolation Products

There are a series of isolation products manufactured by Mason Industries. We'll take a look at how to use them properly in Chapter 10.

Figure 4.29 shows NPS Neoprene Partition Supports. Just as you can float slabs and ceilings, you can also float walls. These partition supports provide isolation from the existing slab to minimize slab transmissions into the wall structure.

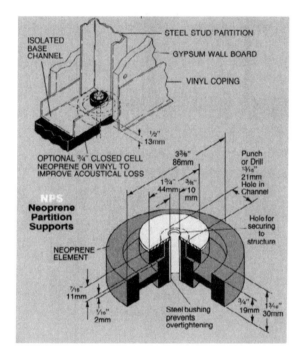

Figure 4.29 Partition supports.

Figure 4.30 shows DNSB wall braces. Sway braces prevent buckling or overturning of tall or extremely long walls.

Figure 4.31 shows WIC and WCL sway braces. These products are ideal for lateral wall support where needed, while still maintaining isolation from the existing structure.

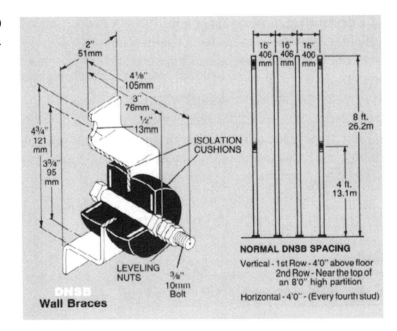

Figure 4.30
Wall braces.

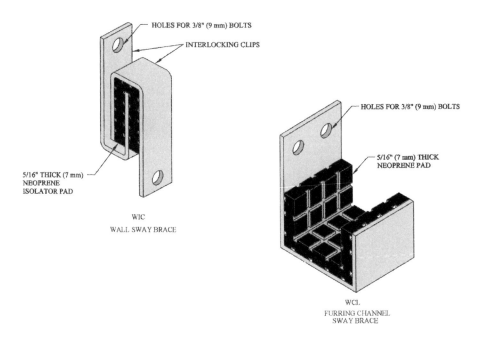

Figure 4.31
Sway braces.

Damping Systems

In a constrained layer damping (CLD) system, a damping material is sandwiched between two other (usually stiff/rigid) materials. Damping occurs when the viscoelastic center of the "sandwich" is sheared (see Figure 4.32).

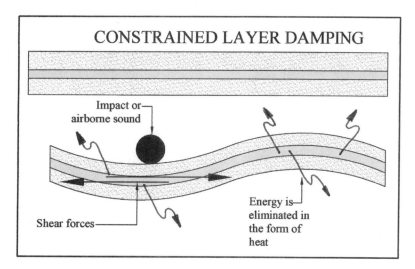

Figure 4.32
Constrained layer damping and how it works.

When bent, shear forces pull and stretch the damping material. Energy is lost when the damping material is sheared. The vibration energy is not isolated, but rather it's destroyed and converted to heat at a rate defined by the efficiency of the damping material in any given system.

This is totally different than using an adhesive to glue multiple pieces of material together. When two pieces of drywall (for example) are joined together through the use of conventional construction adhesives, they bond, forming (essentially) one thicker piece of material. The effective isolation created from this type of assembly is less (as a whole) than the sum of the individual sheet themselves.

Thus, by gluing sheets together (using standard adhesives), you decrease isolation.

But with damping systems, the bonding actions that take place with conventional adhesives never occur, and the isolation is (potentially) greater than the sum of the two sheets themselves.

There are a large number of damping systems on the market: sheet-loaded vinyl, polymeric materials incorporating mineral fillers, etc. But the only one that has caught my attention (providing not only greater isolation at low frequencies than standard drywall, while maintaining a cost performance that's reasonable) is a product called Green Glue, which is manufactured by the Green Glue Company.

Is Green Glue more effective than just adding additional sheets of drywall? Well, if you remember Mass Law—each doubling of mass theoretically adds 6dB of additional isolation. But Mass Law stops at the drywall itself (in this case), so when it comes to a complete wall assembly, reality is closer to 4–5dB. So if you have two sheets on each side of a wall, the next step is four sheets on each side for an averaged increase of a maximum of 5dB.

Figure 4.33 is a comparison of four wall assemblies, three of which were tested at Orfield Labs[3] for the Green Glue Company. The first is a simple single-stud wall with one layer of drywall on each side. The second is a single-stud wall with two layers of ⅝" drywall on each side. The third is a single-stud wall with two layers of drywall on each side, with Green Glue sandwiched between the drywall on both sides. The fourth wall is an estimated performance for a single-stud wall with four layers of drywall on each side. Finally, at the bottom of the sheet, you see a chart line that shows you what Mass Law will gain you for each doubling of mass with this wall assembly, just for comparison's sake.

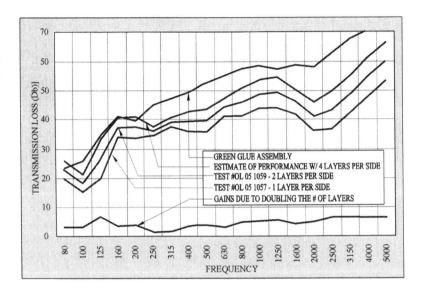

Figure 4.33
Isolation values for various wall types compared to a Green Glue assembly.

Note the increase in performance for the wall with only two layers of drywall on each face by adding Green Glue to the assembly. Note as well the superior performance of that wall to the wall with four layers of drywall and no Green Glue. Although they are close in the 125 to 200Hz range, the Green Glue wall is far superior over 200Hz, as well as around 100Hz.

This is a product well worth considering if weight is a consideration in your structural analysis. It's a very effective means of increasing the ceiling floor assembly without the added weight that you have with strictly a mass approach to your construction.

Whether or not you need this product is something you have to decide. This decision (like everything else) should be based on a careful analysis of needs (meaning how much isolation you really need to provide). If you are a drummer playing light jazz with a set of 7b drumsticks, you can probably handle your isolation needs with some decent isolated construction and a couple of layers of drywall. If (on the other hand) you have a band playing heavy metal at full volume, that would not come even close to providing any reasonable amount of isolation.

Only you can make that determination.

Optional Systems

A system that I have employed with some success since the first edition of this book is a floating wall/ceiling system with a modified floating deck assembly.

The walls are constructed using Mason Industries ND Isolators (see Figure 4.34). After constructing your walls, you can add an isolating floor assembly using 1" or 2" of 703 rigid insulation with two layers of plywood covering that (your new plywood deck).

Although this is not a true floating deck, it does have its benefits. A big plus for this system is that is creates a very effective barrier to impact sound transmission to the room below your floor.

Another benefit is that it does not create a drum head, which is due to the plywood resting completely on a bed of rigid insulation, which effectively damps the surface.

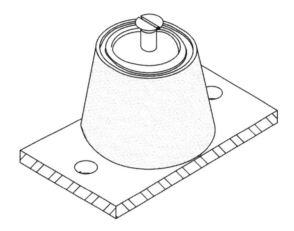

Figure 4.34 ND isolator.

Figure 4.35 shows a section through this assembly. We will examine how to detail and calculate for these supports in Chapter 10.

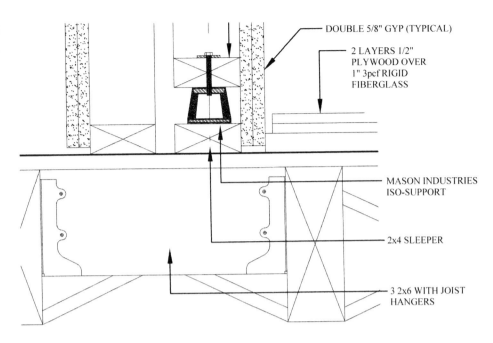

Figure 4.35
Section through wall.

Endnotes

[1] Reprinted with permission of Mason Industries.

[2] Reprinted with permission of Mason Industries.

[3] Orfield Labs, LLC, 2709 E. 25th Street, Mineapolis, MN 55406. Contact: Mr. Steven J. Orfield, Phone: 612-721-2455.

CHAPTER 5

Window and Door Construction

So you understand about Mass Law, about isolation construction techniques, and about how important it is to seal up each and every square inch of space in your walls, floor, and ceilings. You know that where air goes—sound goes. Next, you're going to cut big gaping holes in these assemblies to put in huge windows and doors.

Yup, it's a fact of life—you need to be able to enter and leave the room. You also might want to be able to see the people whom you're recording. On the subject of seeing the people you're recording, if this is a combination room, then you obviously won't need a window. (This is the case in my studio.) You also have the option, in a multiroom studio, of providing video feeds to other rooms versus windows. My preference, however, would be a window in that case. There is something about being able to see (face to face) the people you're recording that doesn't translate to a video screen.

In this chapter, we'll look at exactly how you go about putting these assemblies in while still maintaining your isolation. We'll look at what you need to do to keep air from moving, through the use of special gasketing assemblies. We'll figure out how to improve the quality of standard doors to get that extra isolation you need. We aren't going to do a step-by-step construction of these assemblies in this chapter, because that will take place in the next chapter. Rather, we are going to look at this from the perspective of the assembly as a whole, just to understand how it works as a unit, and which parts and pieces you will need when you finally do build it.

Glass

Because not only the windows, but often the doors will have glass in them, let's begin by taking a look at different types of glass, and the advantages they have (or don't) in sound isolation.

Float Glass

Float glass (annealed glass) is the typical glass you are used to seeing in your windows at home. Most glass today is produced by the float process. The raw materials (primarily silica sand, soda ash, and limestone) are weighed, mixed, and conveyed to a melting furnace. Molten glass flows continuously from the furnace onto a bath of molten tin, where a continuous "ribbon" is formed. The glass floats on the tin and is pulled and stretched to the desired thickness and gradually cooled until it starts to solidify. The glass ribbon is then lifted out of the tin bath onto rollers and conveyed through an annealing lehr.

A *lehr* is a long oven that glass moves through on a conveyor belt. This allows for gradual cooling, which slowly relieves the stresses within the glass and allows the glass to anneal properly. The glass leaving the furnace passes between two metallic rolls that gives it the desired thickness. The glass is then slowly cooled until the glass exits the lehr, slightly above room temperature. At this point, the glass is flat and has virtually parallel surfaces. The glass is then cut, sized, and packaged. The standard specification for flat float glass is ASTM C-1036.

Float glass has an advantage over tempered glass only due to its ability to be cut on-site. It's readily available from any of your local glass shops in thicknesses up to 1". No, I am not referring in this case to an insulated unit—that is the thickness of a single piece of glass. Float glass is the least effective glazing when considering acoustic isolation.

The last sentence in the preceding paragraph is said with one caveat, which is that although standard plate glass is the least effective isolator, there are times when it may be all that you need to maintain the level of isolation in your wall assembly, at which point it becomes a cost-effective solution to your glazing needs. We will examine this in greater detail later in this chapter.

Heat-Strengthened or Tempered Glass

Tempered glass is produced by taking an additional step in the manufacturing process of floating glass. The basic principle employed in the heat-treating process is to create an initial condition of surface and edge compression. This condition is achieved by first heating the glass and then cooling the surfaces rapidly, which leaves the center glass thickness relatively hot compared to the surfaces. As the center thickness then cools, it forces the surfaces and edges into compression. Wind pressure, impact, thermal stresses, or other applied loads must first overcome this compression before there is any possibility of fracture. Understand though, a piece of tempered glass can be completely destroyed by a single nick on the edge that would never affect a standard piece of float glass. I had an insulated tempered glass unit literally explode in my face once while installing a piece of trim around a custom window opening. A 4d finish nail just barely nicked the edge of the glass, which was enough damage to the glass to force it to break into pieces.

Tempered glass is what you normally see installed in doors, door sidelights, sliding glass doors, etc. This glass is required as a safety means for glazing installed within 18" of a floor (this is a starting point—it gets more involved and has to do with pane size, mullion locations, etc.), for glass used in doorway passages, and for any glazing immediately adjacent to those passages. This glass is considered to be "safe," both due to its strength and its ability to break with a unique fracture pattern. Its strength, which effectively resists wind pressure and impact, provides safety in most applications. Also, when fully tempered glass does break, the glass fractures into small, relatively harmless fragments. This phenomenon (referred to as "dicing") reduces the chance of injury. Fully tempered glass is considered a safety glazing material if it is manufactured in accordance with ANSI Z97.1.

Tempered glass is superior to float glass when acoustic isolation is a consideration; however, a disadvantage is that it cannot be cut once the tempering process is complete.

Laminate Glass

A French chemist, Edouard Benedictus, invented laminated glass in 1909 and named it "Triplex." The process bonds two sheets of glass using a sheet of plastic, thereby producing a safety glass. Laminated glass is something you would normally see used for automobile windshields, but it is readily available in flat plate form as well. From a safety point of view, this glass is superior to either standard annealed or tempered. Modern laminated glass panels are typically manufactured using a PVB interlayer. The manufacturing process for this glazing should adhere to ASTM C-1172.

From an acoustical point of view, the big advantage of laminate glass is that the plastic inner layer (or laminate) provides a significant amount of internal structural damping for the glass. This damping effect has a major impact on the sound transmission properties of glass at high frequencies, especially near its critical frequency.

As mentioned earlier, the critical frequency is the acoustic frequency at which the wavelength of bending waves in a surface equals the wavelength of sound in air. At frequencies in the vicinity of the critical frequency, sound waves will pass through the glass much more readily than at other frequencies. As discussed in Chapter 3, "Isolation Techniques—Understanding the Concepts," this effect (reduced sound isolation in the region of the critical frequency) is called the *coincidence effect*.

Critical frequencies of glass are dependent on the glass thickness. Thicker glass will have a lower critical frequency than thinner glass. For example, 1/8" thick float glass has a critical frequency of 4,800Hz, while 1/4" thick float has a critical frequency of 2,400Hz, and 1/2" thick float has a critical frequency of 1,200Hz. Note that each doubling of thickness cuts the critical frequency in half.

Lab tests have proven that the isolation value of laminated glass in the coincidence region is greater than standard float or tempered glass. Thus, laminated glass provides better sound control than regular glass of the same total thickness. Just keep in mind that the benefit is really in the critical frequency range of the coincidence effect.

Plexiglass

Plexiglass is a plastic sheet manufactured for a wide variety of uses. However, none of those uses includes sound isolation. This material just does not have the qualities you need. Remember that you need mass equal to the mass of the wall sheathing, stiffness to control lower frequencies, and a product that is going to hold up well over the lifetime of your studio (hopefully a long time). Plexiglass tends to scratch easily, and it can quickly become tiring to look through once this begins to happen.

So use this product where it's designed to be used, and use glass in your windows and doors.

Window Frame Construction and Isolating Techniques

The window is the easier of the two openings to deal with (which is why I began with it). If you have "room within a room" construction, you just determine what the mass is on your wall assembly and then match it with the window. Simple, isn't it?

Well, maybe not quite that simple. There are some construction techniques that you will have to use.

Window Frames and Trims

In wall construction, it has been proven (through lab testing) that a heavy wooden window frame that runs continuously through the opening does not essentially weaken the TL value of a wall; however, if you absolutely want the "maximum benefit," your best bet is to use a separated wood frame assembly. Light-gauge steel frames (such as the type you would normally see in commercial applications for doors and windows) are a different story, though. So make it a point to stick with wood when building your window frames.

Some of the features of these assemblies will be identical, regardless of the wall type. I will make it a point to discuss only the differences in the assemblies, and I will try not to bore you by being redundant.

Figure 5.1 indicates a 5/4 stock window frame (1 1/4" nominal-sized wood) in a 2x6 wood-framed single wall assembly. It utilizes outside window stops made from standard 3/4" 1/4-round trim stock, a center stop of 1"x6" pine (ripped to fit) with 1/2" and 3/4" glass. The exterior trims are just simple 1"x4" pine.

Chapter 5 Window and Door Construction

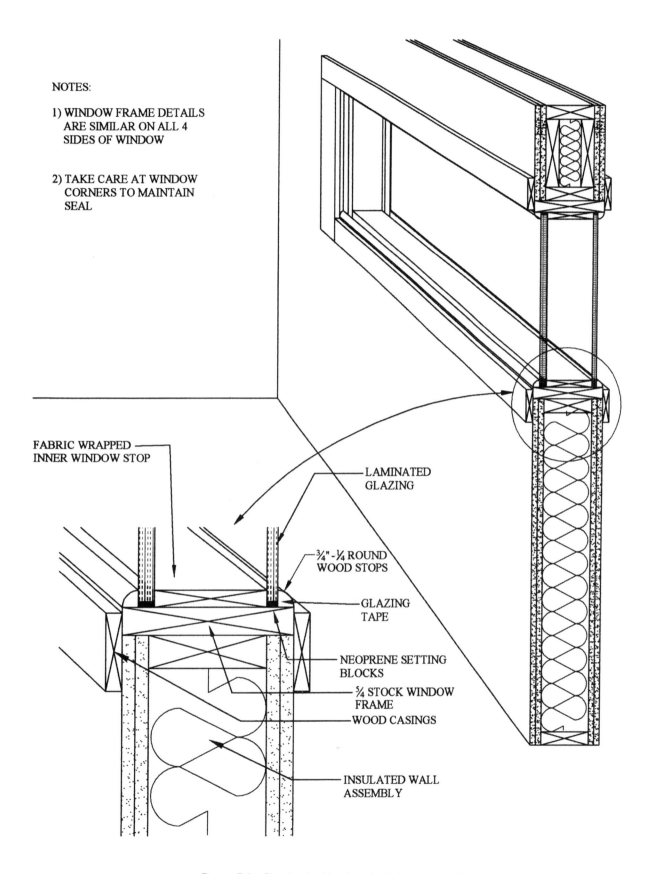

Figure 5.1 Simple double-glazed window assembly.

97

I like to take the inner stop and cover it with black felt, which adds a nice touch. Alternately, you could use a different fabric, or you could finish the wood naturally. It doesn't make a difference from an acoustical point of view—just a professional-looking finished product.

Note the use of neoprene "setting blocks" to set the window on when installing it. These blocks help to maintain the window line in the opening, as well as allow caulking below the window to create an airtight seal. Another piece of the seal assembly for this window is the use of glazing tape for both the inside edge of the window trim and the exterior. Glazing tapes are standard products that can be purchased from any local glass company. I like to use the Extru-Seal manufactured by Precora Corporation for my windows. It's available in $1/16"$ and $1/8"$ thicknesses and $3/8"$ to $1/2"$ widths. Black and gray are the only colors, but I have never used anything other than black. Butyl caulks are perfect for the window perimeter, and they can be purchased at any local lumber center.

In staggered 2x6 stud construction, there is really no easy way with this wall type to isolate the frame from the separate stud assemblies, but when you take into account the coupling of the two wall faces with the top and bottom plate assemblies, you don't really lose a lot of isolation by continuing that "weak point" through the window assembly. However, if you build a staggered stud wall using 2x8 top and bottom plates and 2x4 studs, you can create an isolated frame by following the details in Figure 5.2. This window assembly is typical for an isolated wall, but it can also be used for a staggered stud wall. Care must be taken to ensure that the two frame faces never connect with one another. This can be accomplished by simply ripping the window framing plates to a width that ensures isolation (if need be).

Figure 5.2 is a typical window assembly with separate wall frames. Note how the frames for each window are totally isolated from one another through the use of a rubber seal surrounding the opening between the window frames. (I use this to keep dust from settling into the window opening over time and also to increase the effect of the air spring within the window panels themselves.) The areas between the separate inner window stops are then tightly fitted with rigid fiberglass insulation, and the finished layer is fabric wrapped. This gives a nice finish, and neither the rubber wrap nor the fiberglass can transmit sound from one wall face to the other. Another nice feature about this detail is that the seal gives you a good place to set some bags of desiccant to avoid the possibility of any moisture problems within the window cavity. Although this doesn't happen often, I have had people contact me to ask my assistance in this regard, so it can occur.

Chapter 5 Window and Door Construction

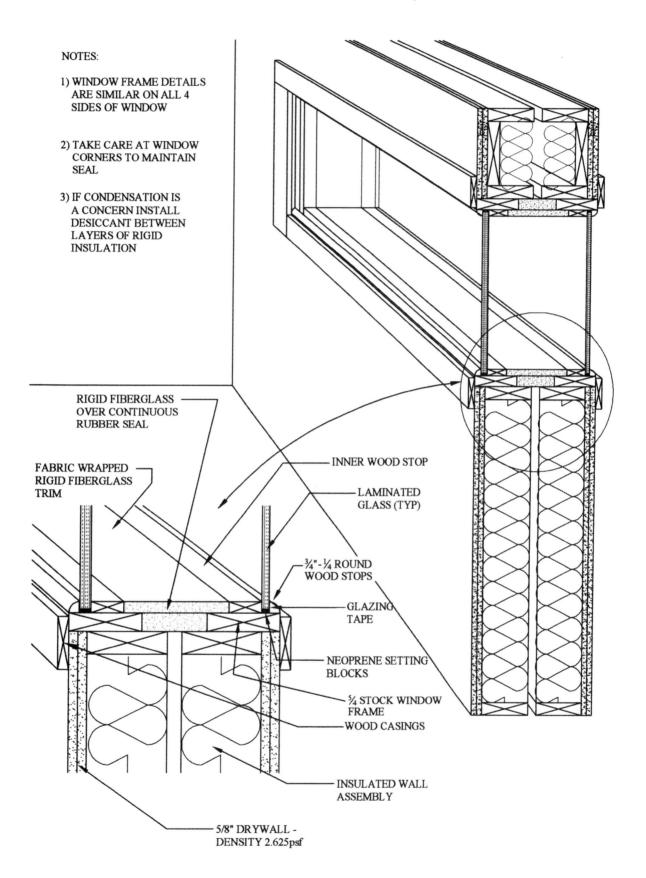

Figure 5.2 Isolated frame double-glazed window.

Glass Thickness

By now you've noticed that I did not indicate any glass thickness in the previous drawings. You should be wondering (at this point), "Exactly how do we go about determining what glass thickness to use in our window openings?"

This is really one of the simpler things to do—just figure out what the weight of your mass on one side of the wall is, match this as a minimum for your thinnest piece of glass, and make sure that you have at least 1/4" in thickness for the glass on the other side of the wall.

Remember the basics—it's all about mass. That doesn't change just because the material changes. For example, if you have two layers of 5/8" drywall on each side of your wall, then the mass = 2×2.625psf (per layer) or 5.25psf of mass (on each side of the wall) for the drywall. Standard plate glass (annealed) weighs 1.64psf for glass 1/8" thick. Thus, a 3/8" piece of annealed weighs 4.92psf (a little light), so you need a piece of 1/2" for one side and 3/4" for the other. So, the glass weights will be 6.56psf and 9.84psf, respectively. Don't worry about the fact that the glass weighs more than the wall—remember the added benefit you get from thicker glass with the lower coincidence effect. That, plus the fact that the coincidence level for each piece of glazing is different, will help you to increase isolation. Because of this, the wall assembly surrounding the window will be your weakest point, and the effective isolation will be as if you had never penetrated the wall at all. Laminated glass is a better isolator than annealed, so if you want to use laminated you could drop those numbers to 3/8" and 5/8".

All of the above assumes that you are using standard construction with typical drywall products. If you are using a product like Green Glue between your drywall layers, then it gets a little more involved. In cases where the construction is outside of the norm, you have to take the time to determine the equivalent mass for your window assembly. Products like Green Glue, mass loaded vinyl (MLV), and special drywall products that include combinations of drywall with special internal damping, all achieve isolation greater than the just the mass of the wall would account for. In those cases, it is important that you determine what the equivalent mass would be and size the glazing accordingly. The easiest way to do this would probably be to contact the manufacturer of the product you are using to verify the actual values you should expect to achieve with your particular construction.

Finally, I want to touch base on the issue of splayed glass (which is when the two pieces of glass are farther apart at the top than they are at the bottom) versus parallel panes of glazing.

Maximum volume equals maximum isolation. So placing the glass more closely together at the bottom decreases the isolation value of the window. But the splaying of the glass also helps you in a few respects—the first of which is that it stops you from seeing reflections in the glass. (By reflections, I mean images, not sound.) It also helps to stop glare from lighting. Avoiding reflected images and lighting glare will make your windows much less tiring to look through. An additional benefit (from my perspective) is that it may help to stop early reflections from the glass reaching your seating position. (By reflections now I'm referring to sound coming from the speakers.) Instead of having sounds reflect from the glass into your face, they will reflect down toward the floor or into the back of your console.

In my mind, the slight loss in isolation from the air spring is made up for in mass. We will always have greater mass than the wall surfaces if we do this right, so what little difference there may be we just live with. In Chapter 1, we looked at the fact that in the real world sometimes there are things that we just live with. This is one of those things that might well be worth living with. I say this because (all other things considered) sloped glass in studio windows is also a great-looking effect.

However, if you want to gain the maximum isolation benefit (at this point the walls should always be the weak point and control the level of isolation), then feel free to install the glass parallel. This is the sort of thing that is really a matter of personal choice, and because it won't affect the outcome to any great degree, I won't push you in any one direction. It's only important to me that you understand enough about the subject to be able to make an informed decision.

Figure 5.3 details a typical splayed window unit. Note that I have now switched from a quarter-round window trim to square edge stops. This glass weighs quite a bit, and the greater the window splay, the more the weight will be carried by the topmost window stop. Don't take chances with small stock if you're heading this route. Make sure that you don't have any surprises coming down the road—use the larger stock.

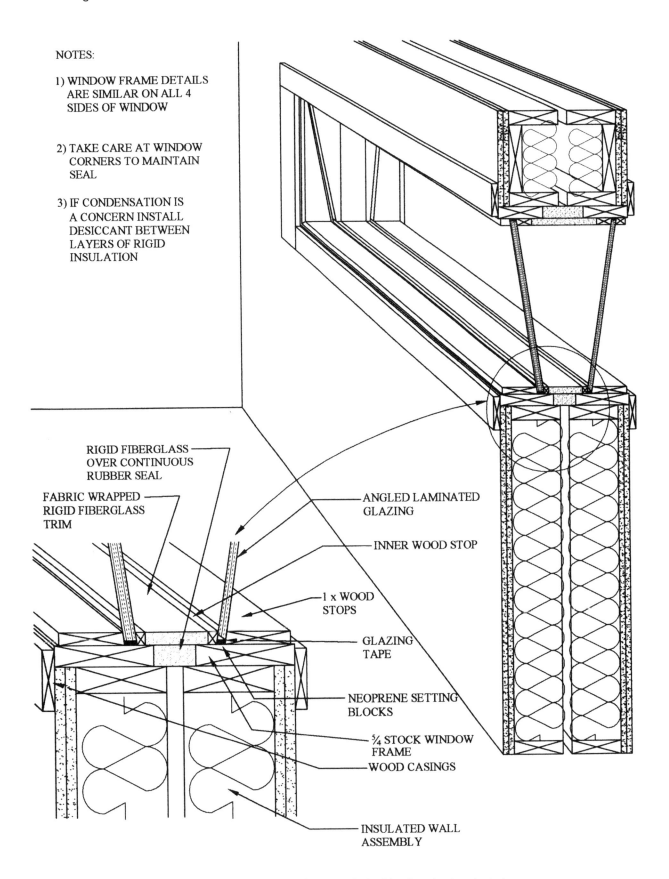

Figure 5.3 Isolated frame with double-glazed splayed window.

Manufactured Window Units

Just to cover all the bases, you can purchase manufactured acoustic window units. Over the years, I've worked successfully with products manufactured by the Overly Door Company. Their window line provides special steel frames with dual-glazed systems, offering acoustic ratings up to STC 55. This rating was achieved with a clear glazing area of 21s.f., using 3/4" and 1/2" laminated glass. The cost for these units is pricey when compared to DIY windows, but if you do not feel up to constructing a window and would rather purchase one, then Overly would be a manufacturer you could not go wrong with. An example of their products can be seen in Figures 5.4 and 5.5. These pictures are from the Quincy Public Access TV Studio located at the Thomas Crane Public Library in Quincy MA.

Figure 5.4[1] Overly Door Co. window.

Figure 5.5 Overly Door Co. window.

Constructing Doors

With door construction, the basics stay the same (at least with the door itself) and that means mass, mass, and more mass. Funny how that theme follows us though this book, isn't it? But, when it comes to the remainder of the door assembly, it's all about stopping the movement of air.

Unlike a window (where you punch a hole through a wall and then seal it up tight again), a door is built with the intent that you are going to continually violate the seal of the wall. So sometimes you want the wall airtight, and sometimes you don't. Here everything becomes an issue of seals—multiple seals, a layering of seals, seals upon seals upon seals, etc.

Do you remember back in Chapter 3 when we looked at that small crack at the bottom of the wall and what that equated to when compared with a hole through the wall? Well, the door is very much like that wall, but with a very big crack at the door bottom and a smaller crack on the remaining three sides. The pain is this—in order for the door to operate properly, you need the door to be fitted to create all those relatively big cracks. Thus, we are forced into those seals mentioned above.

By the time you finish a door, you will be dreaming about seals, because this is going to be make it or break it time for your room. Do this right, and you have it made, but screw this one up, and no matter how good the remainder of your construction is, you just lost it.

Now, in all fairness, the one nice thing about this is that the door is available to take apart without removing all of the assembly (unlike a wall, floor, or ceiling once they're completed), so you can work, then rework, and then rework again if required. But in the end, if it isn't right, neither is your room. So this is a place where you may want to bring some professional help in if you don't feel up to the task.

Another reason to consider this (if you are not a carpenter or a real handy DIY'er) is that the materials for a proper door assembly can get pretty expensive real fast. And if they are accidentally cut short—or otherwise get ruined to the point where they don't work—then replacing them can sure eat into your project budget. Better to keep your extra money for room gear if you aren't up to this one.

OK, now that I've warned you—on to the task at hand.

Door Frame Construction

There are a few different ways of attacking door openings through walls. One is what I call a "super door," which is a modified solid-core door. The second one is through the use of two standard solid-core doors in the same opening. My preference is the super door. Let's take a look at both.

I'll let you know right now that this is one place I don't worry about maintaining the separation of wall assemblies with the frames, even when using totally separated assemblies. When it comes to carrying a door that might weigh well over 300 pounds (if you build them like I do), or even standard solid-core doors, you do not want your door frame attached to a stud that can move over the years. As I noted previously, tests have proven that a through jamb does not effectively lessen the total isolation value of a wall assembly to any great degree. So don't worry about any miniscule amount of isolation you may lose. Just build the frame straight through the cavity.

I would also recommend against using the standard 5/8" jambs that are typically used with pre-hung doors, because the jambs will not hold up against the weight of the "super door" assembly. Often, neither will a standard hinge. Use 5/4 stock lumber for your frames; 3/4" pine is fine for door stops. Then use a variety of special seals to make this baby airtight.

Figure 5.6 is a plan view of a typical door assembly. I begin with a standard solid-core door, and in this case, any manufacturer's door is as good as any other door. As long as it's solid core, don't bother worrying about what the core material is made of.

The door itself is improved by adding mass to the inner face. My preferred method is the use of 8psf sheet lead, which can be purchased at any local masonry supply store in varying widths. This is sandwiched between the door and a piece of 3/4" plywood. The plywood and lead are held back from the door edge 1" (on the two sides and the top) to allow for weather-stripping the door. The door is then hung with heavy-duty hinges, weather stripping is installed, a bottom drop seal (with threshold) is put into place, and then, finally, door latches complete the project.

The final part of this assembly is a door closure. Order a good quality commercial closure for your door. You want this for two reasons, the first of which is safety. Yes, safety. A door with this much mass could break bones, or even remove body parts, if they happened to be between the door and the jamb when it closed. Although you might never have this happen to you (as an adult), the possibility of this happening with a child is greater than you might think. A good commercial door closure has the strength to handle a door of this weight, and it has two distinct closing settings. The first setting is with the door in a wide-open position, and it controls the closing speed of the door to within (roughly) 8" to 10" of the jamb. Then the latch speed takes over, which can be adjusted to complete the closing at a very slow, gentle rate. Not only does this give you some safety, but it will also help your door assembly live longer. By this I mean that the slamming of a door like this will cause screws attaching it at the hinges to loosen up, which should be avoided.

By now you understand that this is going to be an expensive piece of your studio cost. None of these items is inexpensive. But this is one place you do not want to skimp on expenditures. Remember that your entire assemblies are as good as your weakest link.

An alternative to constructing a door like this is to use two doors on opposite sides of your isolated walls, in which case you would construct in accordance with Figure 5.7. Although you save the cost of the sheet lead and cabinet grade plywood, in this case, you will have to purchase two times as much gasketing, drops seals, thresholds, and closures. But this is still an option should you choose to use this technique.

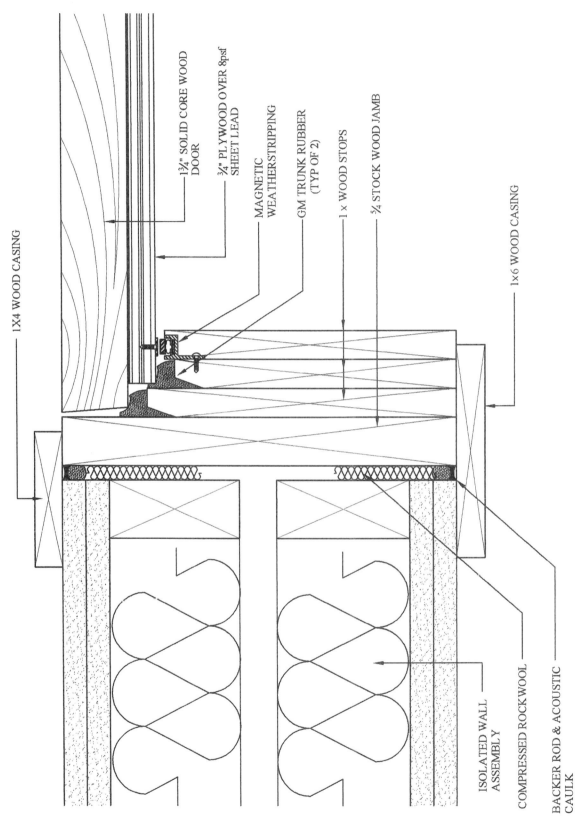

Figure 5.6 Typical door assembly.

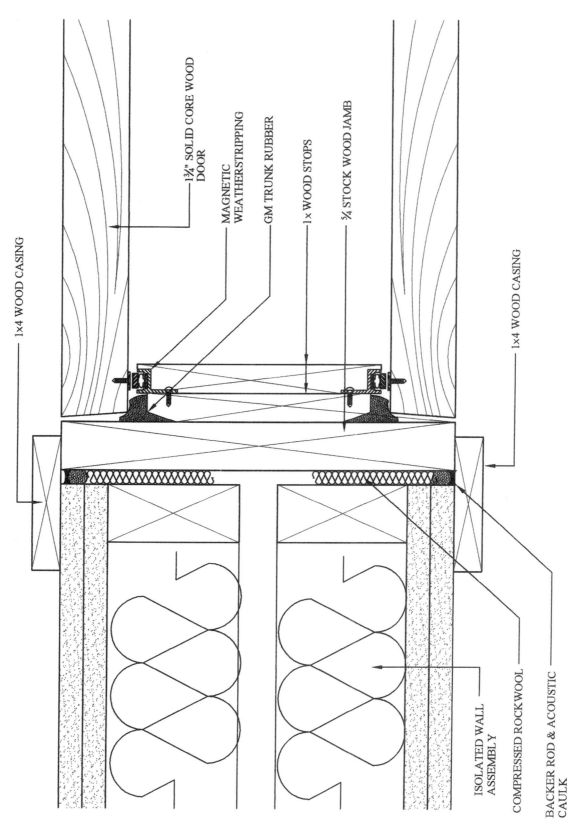

Figure 5.7 Double door assembly.

Windows in Doors

Installing windows in doors is similar to the installation in walls—you want an airspace and airtight construction. However, you become limited in airspace and window thickness, depending on the thickness of the door assembly.

If you use a double door assembly, then you would use the same glass thickness that is used on the walls. For example, one door would have 1/2" glass and the other would be 3/4". Remember that you cannot use float glass in this application.

In a single super door, you can use two separate pieces of glass. The door itself is 1 3/4" thick, and the additional lead (1/8") and 3/4" plywood make a final thickness of 2 5/8". A typical 3068 super door will weigh around 304 pounds, thus about 15.2psf. So a piece of glass 3/4" thick, coupled with a piece 3/8" thick will give comparable mass to the door and allow for a 1 1/2" air gap between panes using flush window stops.

Door Hardware

In order to create airtight assemblies, special hardware is required. I use a soft gasket, manufactured by General Motors for the trunk rubber seal on some of their automobiles, as a part of my assemblies. If you look at Figure 5.8, type "K" is the rubber gasket that works best. Figure 5.9 is a close-up view of that seal.

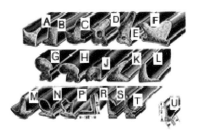

Figure 5.8
General Motors trunk rubber.

Figure 5.9
General Motors type "K" trunk rubber—close view.

I also like seals manufactured by Zero International. They manufacture the following products that I would recommend for door gasketing. Figure 5.10 shows a typical magnetic weatherstrip.

Figure 5.11 is an example of their drop seals for door bottoms.

Figure 5.12 is a threshold that incorporates an additional rubber seal at the door bottom.

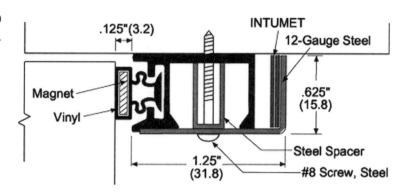

Figure 5.10
Magnetic weatherstrip.

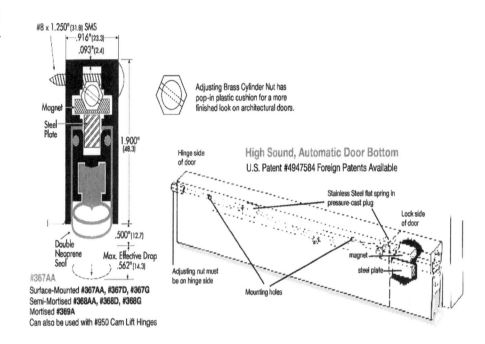

Figure 5.11
Drop seal assembly.

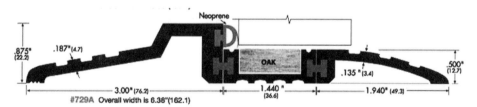

Figure 5.12
Sealing threshold.

Adding Insulating Panels to Door Assemblies

During the testing process for Green Glue, Brian Ravnaas decided (at some point) while taking a wall assembly apart to do a simple test of one-sided assembly. He wanted to find the effect that insulation played in regards to the TL value for something like a door panel.

So they removed the drywall from one side only and experimented with a few different configurations. Figure 5.13 is a chart with the results from those tests.

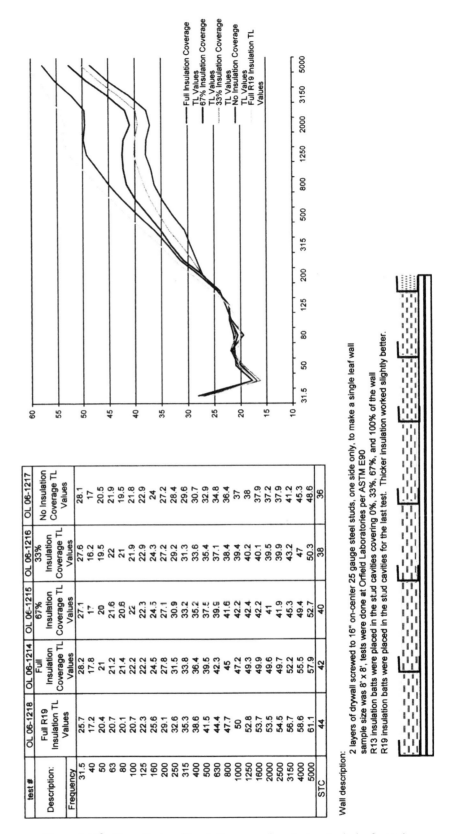

Figure 5.13[2] The effects of insulation paced against a single leaf panel.

The system tested was constructed as follows: two layers of drywall were screwed to 16" on-center 25-gauge steel studs, on one side only, to make a single leaf wall. The sample size was 8'x8', R13 fiberglass batts were placed in the stud cavities covering 0%, 33%, 67%, and 100% of the wall, and R19 insulation batts were placed in the stud cavities for the last test. Thicker insulation worked slightly better. The tests were done at Orfield Laboratories in accordance with ASTM E90.

Although the test results indicate a slight drop in TL value for the lower frequencies, I do not find this anything to worry about. The reason is that a door is a much stiffer assembly (with a lot more mass) than the wall panel they tested, and it will not exhibit the same TL loss at those frequencies as the drywall panel did, perhaps no loss at all.

The weak point with your door assemblies is in high, not low, frequencies. And this is where the tests proved the advantage.

To quote Brian:

"The results speak for themselves. Covering a single leaf wall (like a door) in absorbing material results is a profound improvement in transmission loss, beginning at the frequency where the absorbing material in question becomes very effective. Covering the entire surface is distinctly more effective than covering just a portion of the surface at higher frequencies. For doors, this should prove highly useful, as they tend to be fairly strong at lower frequencies, but weaker at higher frequencies. We anticipate that the frequency at which the absorbing material layer will start to be meaningfully effective will be, as is true for sound absorption, higher with thicker material, etc. So 3" of 703 should work better and lower than 3" of R13, 4" of 703 should be effective to a lower frequency, and more effective overall than 2" of 703."

The Finished Product

Listen up, I urge you to get as creative as you can with room finishes. You are musicians—you have great imaginations, you're creative by nature. Doors (like anything else) do not have to be boring.

Figure 5.14 is a picture of the door entering a studio in NYC (G&E Music). To its left, you can see the door that leads into the iso-booth. This door finish was the brainchild of Gary Everett, a first-class carpenter, and the lead carpenter for that studio build. Gary is also both a musician and Luthier, and has built some guitars with absolutely great sounds.

These people got creative in a fairly small lobby, using a door finish that makes it feel much larger than it really is. Besides which, things like this are conversation pieces. Believe me when I tell you, the people going to this studio will remember those doors for a long time to come.

Figure 5.14[3]
An interesting door finish.

Manufactured Doors

As with windows, should you choose not to construct your own doors, you can purchase manufactured units specifically made for sound isolation. Again, I have had good luck in the past with Overly. They manufacture steel and wood acoustic door assemblies with special hardware, integral gasketing, and isolated steel frames. Their steel doors have isolation values up to STC 55, while the wood units are available with ratings to STC 49. Figure 5.15 is an Overly door with a vision panel.

Chapter 5   Window and Door Construction

Figure 5.15[4]
Overly door.

Only a few more chapters, and we'll begin to look at how this all comes together.

Endnotes

[1] Photography by the Overly Door Company, printed with permission of Overly Door Company.

[2] Data provided by The Green Glue Company, reproduced with permission of Green Glue Company.

[3] Photography by Erik Blicker, printed with permission of Erick Blicker and G&E Music.

[4] Photography by the Overly Door Company, printed with permission of Overly Door Company.

CHAPTER 6

Electrical Considerations

Let's begin this chapter (and the next for that matter) with my explaining that I do not intend to tell you how to wire electrical outlets or install HVAC systems.

Over the next two chapters, I want to explain enough about the subjects of electical system design and HVAC design to help you avoid some problems, allow you to make some decent judgments, and get you ready to deal with the subjects.

If you want to learn how to wire or how to install HVAC systems, there are already some excellent books on the market to help you along, so please buy one of them.

Line Voltage

You need to begin with understanding your electrical service and panel, along with the parts and pieces of your room and gear.

Typical home electrical service will be single phase 230V AC. This consists of a three-line service from the power company into your electrical panel—two "hot" legs of 110V and one neutral (return) leg (see Figure 6.1).

A Comment on Power Company Service

Most schematics for this would indicate the neutral feed flowing in the direction of the service; however, I indicate this as flowing out of the service back to the power company, for that is really its purpose. The neutral returns residual electricity back to the power company's ground.

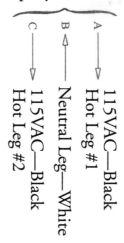

Figure 6.1 Electrical service details.

Most of your use will require a single-pole circuit, which utilizes one of the hot legs, the return, and a ground. On your main panel, the ground bar and neutral bar will be "bonded," which means they are joined together.

One thing that's important for you to understand is that your panel is not built for power leg #1 to feed the left side of the panel, with leg #2 feeding the right side. Rather, power in a panel usually feeds from top to bottom and rotates. So leg #1 are the odd-numbered breakers top to bottom, and leg #2 are the even breakers top to bottom.

Figure 6.2 indicates a typical panel with the main breaker, the neutral buss bar, the ground buss bar, and the sub-breaker configuration. Pay attention to the sub-breaker configuration because this will be important to you when wiring your studio.

One of your goals with studio wiring is to put all of your gear on one leg of the panel and all lighting, such as fans, HVAC equipment, refrigerators, etc., on the second leg. This will help you avoid things like 60Hz motor transmissions from showing up as noise in your recordings.

Although this sounds easy, it may not be the case. This is due to the fact that your panel also needs to have what is called a "balanced load." Picture a hot summer day, all the AC units running, refrigerator running, a load of clothes in the wash, another one in the dryer, hot water heater on, most of the house lights on, someone running a TV, someone else at a computer, and you in your studio with a band. If the total load for your electrical service (at that time) were 170 amps, you would want to draw roughly 85 amps on one leg and the other 85 amps on the second leg. That's a balanced service.

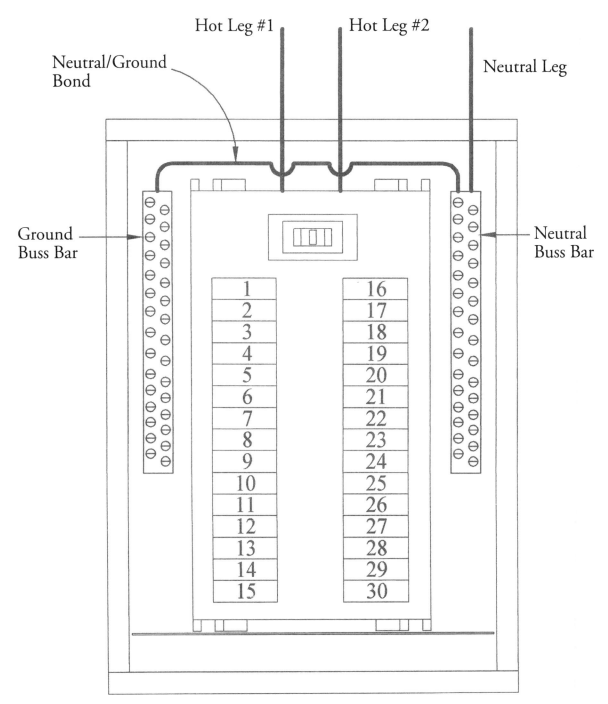

Figure 6.2 Typical electrical panel.

Why? Picture a 200 amp 120/240V residential service with the entire house wired to one phase. Imagine 210 amps of 120V line-to-neutral loads being connected to this phase. This setup will only allow half of the possible amperage load before the main breaker trips because the load on that pole of the breaker exceeds 200 amps. Now, if you take that same load and spread it out over two phases (110 amps per phase), then the main breaker will be fine because the 110 amps is well below each pole's capacity.

Another reason this is desirable is due to the fact that the more balanced the load is on a service, the less current will be carried through the neutral. In a perfectly balanced panel, the current present in the neutral would be 0 amps. This is a safety feature for both the workers of the local electrical company, as well as you or anyone else who might work on your wiring in the future. It's a real pain (potentially a life-threatening pain, in fact) to go to work on a circuit that you've turned off the current to and have current present in the neutral on that circuit that decides you're a good source to ground.

Low Voltage

Low-voltage wiring should be kept as separate as possible from line-voltage sources. In an ideal world, the low-voltage wiring runs low in the room and the line-voltage wiring runs high (or vice versa). If you have to cross low- and line-voltage wiring, then cross them at a 90° angle. Do not run them parallel to one another, or (if you have no choice) try to keep them at least 3' from one another. Take care when running these lines to help minimize noise in your final product.

Figure 6.3 is what you should do if you are putting in a new slab. This requires exacting layout control to make certain every riser winds up inside of walls or in the right location beneath the desk/board. Note the use of steel brackets to hold the conduit in place prior to the concrete pour. To provide power to the desk, take an independent run from a side wall—do not run parallel to the low-voltage runs.

Figure 6.3[1]
Low-voltage conduits.

Electrical Noise

Here's something I hear on a regular basis. "I wired my studio up and everything seems to be wired right, but I have this buzzing sound in my monitors. What did I do wrong?"

Ground Loops

The vast majority of noise issues with sound systems can be traced to ground loops. In order to counter this, you must first understand what is really taking place, and there are a couple of different causes for this problem, often making it even more difficult to track down.

The components of your gear all have their own internal ground, typically referred to as the *audio signal ground*. When you wire up your equipment, it will be interconnected, which can tie the signal grounds of the units together. Ground loops will form when the grounds of separate units also have ties to another source of ground. This could be through the chassis ground via the rack rails, the third prong of the ground on the power cord, etc.

What actually takes place is the formation of a circuit that creates a closed "loop" between multiple units, which then flows through the audio signal ground. This creates a 50–60Hz hum in the sound system.

Another potential source for ground loops is multiple ground sources than can exist in your electrical system service. Electrical service can have several sources to ground because it is a requirement of the NEC that anything in the building with the potential for carrying current be bonded to the electrical systems ground. Quite often these items will also have their own grounds or serve as a ground. Examples of the former would be the steel frame of a building, the reinforcing used in an in-ground swimming pool, the steel frame of an above-ground swimming pool, and steel siding. An example of the latter would be a metallic water service pipe.

All of these either have their own ground or create a ground by virtue of the fact that they create a ground, and all are connected to the ground of the electrical service.

Other sources that can create problems are cable and satellite services, as well as telephone service (all of which will generally have their own small ground rods), and electrical equipment like fan motors, air conditioners, compressors, etc. Figure 6.4 is a simple schematic diagram indicating two circuits with typical grounding.

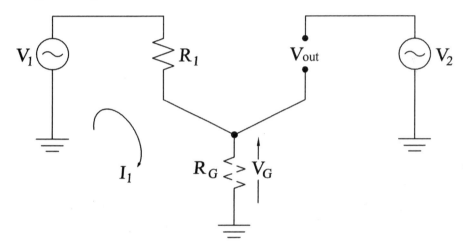

Figure 6.4
Circuit diagram.

The two separate circuits in Figure 6.4 share a common wire connecting them to the ground. In a perfect world, the ground conductor would have no resistance ($R_G = 0$) and the voltage drop (V_G) across it would also be zero. This would have the effect of isolating the two circuits at the point where they connect to one another (a constant ground potential).

In the diagram above, the voltage output from circuit 2 is zero ($V_{OUT} = V_2$.); however, if the ground were to have any significant amount of resistance (for example: a bad connection between the ground rod and grounding wire/clamp) and a current flow (the magnitude of which was I) existed through circuit 1, then a voltage drop would take place across R_G ($V_G = I_1 R_G$) and the ground of the circuits would no longer be at ground potential.

This voltage would actually be applied to circuit 2 (the path of least resistance) and added to the output.

This would be represented as:

$$V_{out} = V_2 - V_G = V_2 - \frac{R_G}{R_G + R_1} V_1$$

The two circuits are no longer isolated from one another, and circuit 1 can introduce interference into the output of circuit 2. If circuit 2 feeds the gear in your room (and circuit 1 has large currents running through it), then you may well have a very audible 50 or 60Hz hum (possibly even harmonics of those frequencies: 100Hz, 120Hz, etc.) coming out of your speakers. This hum would also be evident in your finished recordings.

In addition to that, both circuits would now have voltage (V_G) on the grounded parts (chassis and possibly the case) that might present a potential shock hazard. This would be true even if you turned off the breaker for circuit 2 because the ground cannot be broken between the circuits by turning off a circuit breaker.

Understand that the mere presence of electrical current is not necessarily in and of itself always a cause for hum in a sound system; it is only when the current flows through the audio signal that ground loop noise becomes evident. The voltage that exists is developed between the various pieces of equipment interconnected to create your sound system. Due to the fact that this ends up creating a low-impedance, low-voltage circuit, the current is quite high, and it is this current that is the source of the noise you would hear.

Here's an example I ran across of a ground loop caused by multiple sources to ground that involve two buildings with separate electrical services but a common water service.

A world-class studio we constructed had hum and buzz problems with single-loop guitar coils. The buzz would change as the guitar was moved—from almost totally silent to loud—*really loud*. A test with an oscilloscope indicated a strong 60Hz signal on the ground of a circuit with nothing running on it.

It took almost a month to finally trace the problem. When the building was constructed, the water service to the building was provided from an adjacent building on the property. It was more than enough water to provide our needs for the few bathrooms we had in the new building. However, it was bonded in the original building to the earth ground and then again to the earth ground for our new electrical service. That created a path to two grounds. To solve the problem, we installed a nonmetallic connector (a dielectric union) on the water service between the two buildings, thus breaking the second path to ground.

Simple solution—simple problem—four weeks to figure it out.

Ground Loop Solutions

So how can you go about countering this?

For ground loops caused by multiple ground sources via the electrical service:

- ▶ Remove one of the ground paths, thus converting the system to a single-point ground, in a similar manner to our solution with the water service between the two buildings.

- ▶ Isolate one of the ground paths with an isolation transformer, common mode choke, optical coupler, balanced circuitry, or frequency selective grounding.

In my opinion (as regards the second option listed above), the easiest (and usually a fairly cost-effective) method for doing this would be to use an isolation transformer. An isolation transformer is a device that allows all the desired signals to pass freely, while interrupting ground continuity, hence breaking ground loops.

For ground loops caused by multiple ground sources caused by the interconnection of your equipment, a good start would be to consider using all balanced connections between your equipment. This is considered the best method for interconnecting sound equipment as outlined in the Audio Engineering Society (AES) standards document AES48[2]

I am not going to go any deeper into this subject (ground loops that is) because of how involved this becomes when you really start looking at it closely. What you see above gives you the basics, but is really little more than scratching the surface. All you really need to concern yourself with is developing a good sense of how to properly interconnect your gear, and for that the publication noted above (the AES standard on interconnections "Grounding and EMC practices—Shields of conductors in audio equipment containing active circuitry."is your best bet. You can download a PDF copy at the AES Web site. It is a short (and inexpensive) 13-page document that does not go into a lot of theory, and specifically tells you how (and what) to do, as well as what is not acceptable for audio cabling. The people at AES are some of the more brilliant minds in the world when it comes to these things, so make good use of the work they have already done.

There can be other sources of noise in your gear that have nothing to do with ground loops and can occur even if you install an isolated ground transformer. Creating a quiet electrical system requires careful planning of your service to the studio.

Let's look at some things you can plan (or watch out) for that will help to ensure clean power in your studio.

Isolated Ground Receptacles and Star Grounding

Something you might want to consider when planning the wiring for your studio is the use of hospital-grade, isolated ground receptacles and a star grounding system. This is the most commonly used system for technical grounding. Hospitals use this approach to power distribution as a matter of course. It helps them ensure that their critical equipment will not fail due to dirty electrical circuits. It has a minimum of technical compromises, meets the requirements of equipment grounding, and provides a system that is relatively practical to install, troubleshoot, and maintain.

Figure 6.5 illustrates the basic geometry for a star ground system. Note that the isolated ground wires have been run back to a separate ground panel. The ground buss within that panel is then tied to the earth ground for the building service.

The usual three wires for the receptacles are run normally. Black wire (hot leg) to brass screw, neutral (white wire) to silver, bare ground to box, but then a separate jacketed ground is run back to the ground buss. Typically, this would be run within a conduit (it is required to be protected by the NEC) and should be run in close proximity to the cable used for the receptacle service. Standard junction boxes can be used for this purpose to allow for multiple conductors to be run within one conduit rather than individual ones. The wires can run back to the original ground buss in the main panel, or a separate ground panel with its own ground buss can be used adjacent to the main panel. It's important that this then be tied back to the ground wire running to earth ground on the main panel.

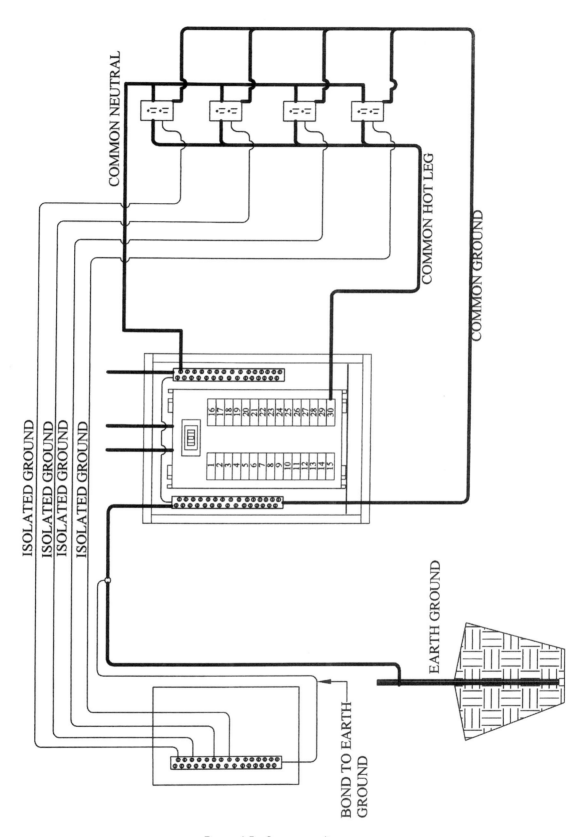

Figure 6.5 Star grounding system.

Lighting

Avoid the use of fluorescent lighting. Although the lighting itself is great and is relatively inexpensive, the ballasts (and yes this includes electronic ballasts) are very noisy.

If at all possible, you should also avoid the use of simple switchable dimmer systems. You can plan your lighting to achieve almost the same effect by using a greater number of lights with smaller wattage bulbs, and you can control intensity through the use of individual switches controlling the lighting symmetrically in pairs.

If you really need (or just want) a dimmer system, stick with good-quality professional systems, not the single slide bar controls that you use for your living room, and understand that you might have to add remotely mounted systems specially designed to get rid of any problems that may crop up down the road.

One quiet source of lighting is LED (Liquid Emitting Diode). I recently had the pleasure of working on a project where we installed this lighting in the public areas, and I will assure you that I was pleased with the quality of light these fixtures produce, much more so than fluorescent lighting.

Although this lighting is pricey—roughly twice the cost of standard fixtures—it is well worth that cost when you consider the savings in operations. This is both from the view of electric company fees and the savings in cooling costs. These units put out very little heat.

These are low-voltage lighting systems, which are available to retrofit existing lighting circuits that have live voltage sources for power. A great reason to use LED lighting is due to the fact that LED lights do not emit radio frequencies.

One of the nice things about the new LED fixtures is that not only are they dimmable, but also a single fixture can accomplish both dimming and color changes using modern programmable control systems. Thus, you can preprogram a series of mood-lighting schemes, with various settings, and then using a single switch, you can change to whatever lighting scheme you desire. A downside to LED lighting is that it is more directional than other forms of lighting you might typically use, and as such, it may require (depending on ceiling heights and room layout) more fixtures for general room lighting than standard incandescent or fluorescent lighting.

I have used lighting systems manufactured by Lutron Electronics, Inc., successfully in the past; however, care must still be taken. The following information comes from Lutron.

What Is Radio Frequency Interference (RFI)?

RFI is a buzzing noise, which may occur in some audio and radio equipment when solid-state dimmers are used nearby. Although every Lutron dimmer contains a filter to suppress RFI, additional filtering may be required in some applications. Typical examples of RFI-sensitive equipment are AM radios, stereo sound systems, broadcasting equipment, intercom systems, public address systems, and wireless telephones.

RFI can be transmitted in two ways: radiated and conducted.

Note: These suggestions will help minimize RFI; however, they do not guarantee that RFI will be completely eliminated.

Radiated RFI

Any sensitive equipment that is in close proximity to dimming equipment can pick up the RFI and generate noise into its system.

The following are three possible ways to minimize radiated RFI:

- ▶ Physically separate the RFI-sensitive equipment from the dimmer and its wiring.
- ▶ Run dimmer wiring in its own metal conduit.
- ▶ Use a lamp debuzzing coil (available from Lutron) to filter the RFI.

Conducted RFI

In some cases, RFI is conducted through the building wiring and directly into the AC power supply of the sensitive equipment.

To minimize conducted RFI, follow these guidelines:

- ▶ Feed sensitive equipment from a circuit without a dimmer on it.
- ▶ Add a power-line filter to the sensitive equipment.
- ▶ Add shielded wire for all microphones and input cables. Also, use low-impedance balanced microphone cables, which are less susceptible to interference than high-impedance types.
- ▶ Make sure that all the equipment is grounded. Connect all shields to the ground at one point. Ground lighting fixture metal housings properly.
- ▶ Use a lamp debuzzing coil (available from Lutron) to filter the RFI.

Lamp debuzzing coils (LDCs) are the most effective way to reduce RFI. One LDC is required for each dimmer. Select the LDC according to the connected lighting load. The LDCs may be wired in series on either the line side or the load side of the dimmer. For maximum RFI suppression, keep the wiring between the LDC and the dimmer as short as possible.

Since the LDC itself makes an audible buzz, mount it in a location where the noise will not be objectionable (for example, an electrical closet, a basement, or above a drop ceiling). LDCs are designed to mount easily onto a standard 4"x4" junction box. They are UL listed and thermally protected.[3]

The How's and Why's of Lighting Noise

Why does it work this way? Why do lights make noise when dimmed?

Solid-state dimmers are electronic switches that rapidly turn the current feeding a lighting circuit on and off, 120 times per second, to produce the dimming effect. This rapid switching can cause incandescent lamp filaments to vibrate, resulting in a buzzing noise. Lamp buzz is generally noisiest at mid-range (50%) dimming level. Some lamps are noisier than others when dimmed, depending on the physical characteristics of the lamp filament. Lamps of higher wattage (100W and above) tend to produce a louder buzz. Therefore, use a lower wattage lamp whenever possible to reduce lamp buzz.

How does a lamp debuzzing coil (LDC) work? When an LDC is wired in series with the dimmer, it slows down the inrush of current during the rapid switching cycle of the dimmer. As the current inrush is slowed down, the lamp filament vibration and lamp buzz are reduced. This helps to quiet not only the lamp itself, but also the amount of RFI transmitted through the wiring system.

Diagnosing and Troubleshooting Problems

To accurately determine the correct solution to a problem, you have to find and isolate it first. For example, if you simply start flailing away, swapping gear and cables and everything all at once, you may never know what actually caused (or fixed) the problem. In addition, you may end up making more and more work since you are expending energy in areas that don't have any effect on the problem at hand.

Start simple. Troubleshooting ground loops involves taking things in order and checking a few basic, common elements to see if the problem is simple or complex. For example, if adjusting the volume on your processor/receiver does not alter the hum level, then the problem must be occurring *after* that point. If it occurred prior, then the receiver/processor would typically alter the overall level of noise. Make sense?

Work in the following methodical manner. Begin by taking one piece of the chain out of the loop at a time. If the noise suddenly disappears when you disconnect a compressor (for example), then the problem lies there, either within the compressor itself or within its power supply.

Another example would be to check your mic or equipment cords. Quite often a bad cord will cause a hum. Begin (again) by checking one cord at a time. Pulling out a bunch of cords all at once might well solve your problem, but it won't let you know which of those cords was the real problem.

Some Common Problems

First, check to see if you have a power cord or an outlet in the wall that is worn out and will not grip. If the hot/neutral/ground prongs on the plug are making intermittent or light contact with the tang on the inside of the outlet, it can cause a hum through the system. The best solution for this is to replace the cord with a new one or the outlet with an industrial version. Industrial outlets have better gripping and will hold power cables more securely.

Second, check the polarity of the outlet because it may be wired backward. You can get a polarity checker for about $5. This is one of the first things you may want to check after you finish wiring your studio. Check this even if you have a professional electrician do the wiring for you. You would be amazed the number of times I find problems with polarity in an outlet on projects wired by licensed electricians. Remember—they're human—they make mistakes.

Third, light dimmers, fluorescent lamps, and other appliances that share the same circuit or common ground with your equipment can cause hums.

Fixing the Problem

If your problem relates to any of the three items above, then change the cords, fix the polarity, or put those lights on a different leg of the service.

Understand, though, that troubleshooting all of the possible sources for hums and buzzes could easily cost thousands of dollars of a professional's time.

"OK, so I've done everything right, still have noise, and found out that the problem stems from the power company itself. They've told me that the power is clean enough for residential use, and there is nothing they can (or will) do to fix it. Where do I go from here?"

If this is the case, then you will have to consider the use of an isolation transformer for your gear.

Tripp Lite is a manufacturer of small isolation transformers. Costs for their units range from around $135 for a two-outlet, 250-watt transformer (see Figure 6.6) to about $343 for a four-outlet 1,000-watt hospital-grade transformer (see Figure 6.7).

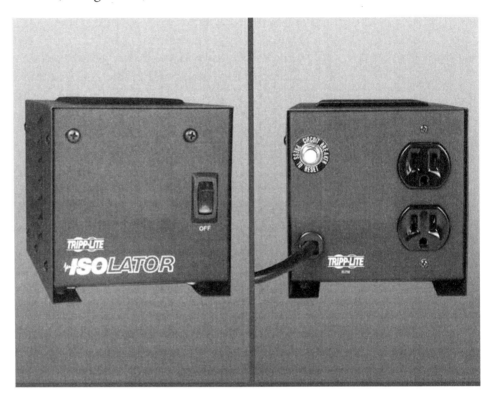

Figure 6.6
250-watt isolated transformer.

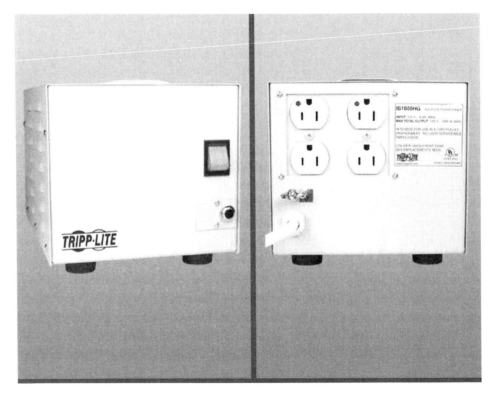

Figure 6.7
1,000-watt isolated transformer (hospital grade).

Tripp Lite uses a design incorporating a Faraday Shield for effective isolation. The Faraday Cage was invented by Michael Faraday (1791–1867). In 1836, Faraday built a room, coated it with metal foil, and allowed high-voltage discharges from an electrostatic generator to strike the outside of the room. He then used an electroscope to show that there was no excess electric charge on the inside of the room's walls and volume.

The same principle is used for shielding of electrostatic RFI from any of the higher quality isolation transformers on the market. Use of these units is simple. You purchase a unit, plug it into your power supply, and then plug your gear into the available outlets. Hopefully, it won't come to this, but these units are always available solutions should the need arise.

Endnotes

[1] Photography by Max Dearing, printed with permission of Max Dearing and Dark Pine Studios.

[2] Audio Engineering Society (AES) AES48-2005: AES standard on interconnections "Grounding and EMC practices – Shields of conductors in audio equipment containing active circuitry." Audio Engineering Society, New York, 2010.

[3] Reprinted with permission of Lutron Electronics, Inc.

CHAPTER 7

HVAC Design Concepts

This is probably one of the most missed items when it comes to home studios. I am constantly bombarded by people coming to ask how they can fix their room. The work's all done, it's airtight, beautiful, sounds great, and they suddenly realize that they need fresh air, air conditioning, heat, the usual amenities.

I shake my head in wonder as I think about how many people miss this so very basic requirement. Re-work is always a lot more expensive than doing it right the first time. So save your money, keep your sanity, and plan for this now.

Again, remember, this chapter isn't going to teach you how to install these systems. It won't tell you how to sweat a copper joint or how to braze compression lines for coolants. It will give you the knowledge of the design concepts with enough depth to discuss your needs with professionals intelligently. Beyond this, you need professionals for the final design and installation on HVAC systems. There are too many different factors that come into play with HVAC systems to cover them all in this book. Heck, you could fill an entire book by itself just going over the basics in design, but if you did that, you wouldn't ever get to the part about building your studio.

Getting Started

First and foremost, understand that you have a super-insulated structure to deal with. All of this work that you're going to do to keep sound in (and out) also helps tremendously to reduce thermal transmissions through the structure. Thus, the design *cannot* be based on the same criteria as your home.

Next, if you have separate control and sound rooms, then you have two very different requirements for these. The control room (for example) will typically be smaller in volume than your sound room, but will generally have more gear producing heat that needs to be dealt with. The sound room (on the other hand) will usually have greater heat and moisture from the human body than the control room, and this too needs to be dealt with.

So it isn't as simple as it might appear. Let's begin with your control room.

Room Design Criteria

The beginning of this is an assessment (on your part) as to the equipment you have (and will have) within this space. Yep, you have to reasonably anticipate what's coming in the future. The reason for this is that all of this equipment gives off heat when it's working, and you need to deal with that heat. Once you have this figured out, make up a list and include, as a part of that list, what the total wattage is for each and every piece of gear.

If you are going to install an isolation transformer in a location that will transfer heat into the room, you should include this in your calculations. Again, provide the total wattage for the unit. The same would go for a computer, UPS, and so on.

Lighting adds quite a bit of load to room cooling requirements, so a simple calculation based on the area of the room should be enough to cover this, unless you have some plans for very heavy lighting loads.

Next, you have to decide just how many people are going to be inside that room besides yourself. Once again, wattage is the data required for design. Yes, the heat that people give off can be calculated as wattage, you'll learn how it's done.

Finally, fresh air is taken into account.

All of this information is critical to proper HVAC design.

Btu Output

Let's take a look at how this info is utilized.

One Btu (British thermal unit) is the amount of heat required to increase the temperature of one pound (one pint) of water by one degree F. This is roughly equal to the heat produced by one standard wooden match.

Thus, to bring one pound of water from freezing (32°F) to the state of boiling (212°F) requires 180Btu (i.e., 212°F- 32°F = 180°F = 180Btu).

To convert wattage into Btu output, you multiply watts times 3.4129. To convert Btu's to watts, simply multiply times 0.2930711.

The People Factor

Calculating what people contribute to a room gets a little more involved.

To accurately calculate the Btu output of a person, you first have to determine their BSA (Body Surface Area). To determine this, you could use one of several mathematical models available. I prefer the Mosteller[1] formula, but other popular methods exist. My reason for preferring this formula over the others only has to do with the ease of use this one presents. The calculations can be performed with a simple handheld calculator. The Mosteller formula is:

BSA (m²) = ([Height(cm) x Weight(kg)]/ 3600)* .5

A person 5'-10" tall (177.8cm) weighing 180.78 pounds (82kg) would have a BSA of 2.02594m²

BSA (m²) = ([177.8cm × 180.78kg]/ 3600)* .5 = 2.02594m²

To convert meters² to feet², simply multiply it times 10.7639104. So your friend above has a BSA of 21.807 feet², and let's call him Mr. "A."

OK, now that you've dragged yourself through all of that, where do you go from here?

Well, if you look at Figure 7.1 you'll see a set of values for the metabolic rate of people doing various tasks. The values are indicated as "met units." One met unit equals the output of one square meter of skin.

Activity	Metabolic Rate (met units)
Reclining	0.8
Seated, quietly	1.0
Sedentary activity (office, dwelling, lab, school)	1.2
Standing, relaxed	1.2
Light activity, standing (shopping, lab, light industry)	1.6
Medium activity, standing (shop assistant, domestic work, machine work)	2.0
High activity (heavy machine work, garage work)	3.0

One met = 58.2 W/m2 = 18.4 Btu/h ft2

Figure 7.1 Metabolic rate chart.

This value (for 1 met unit) is 58.2 watts per meter squared or 18.46Btu per hour per square foot (58.2 watts * 3.4129)/10.76 = 18.45Btu/h f²). In an office setting, you would multiply this times 1.2 for a total of 69.84 watts (22.14Btu/h f²).

So, using the example above, Mr. "A" will produce (in a typical office setting) 482.81Btu/h or 141.48 watts/h (22.14Btu/h x 21.807 feet²). You're going to use the data from Mr. "A" to represent the "typical adult" throughout these exercises.

Room Calculations

Let's create a 16x20 control room and fill it with the following equipment:

- ▶ One 24-Channel Mixing Board with a 250 W power supply
- ▶ One Stereo Vacuum Tube Power Amplifier, 500 W
- ▶ Two Dual-Channel Tube Mic Preamps, 35 W each
- ▶ Four 4-Channel Stereo Compressor Limiters, 35 W each
- ▶ One Stereo Tube Limiter and Mic Preamp, 75 W

- One Stereo EQ, 72 W
- Six Effect Units, 16 W each
- One 6-Channel Headphone System, 210 W
- One 24-Channel Digital Recorder, 50 W
- One 2-Track ¼"x7" Stereo Reel to Reel, 30 W

That's 1,493 watts of gear in a 320 s.f. room that will hold a maximum of five people during mix-down. You decided you want clean power, so you're using a 500-watt isolation transformer that will sit beside your desk.

If you're building a studio with any serious sound-isolating capabilities, then you don't have to bother taking passive solar gain into consideration, nor is thermal loss of any real consequence. There is, however, one other item to deal with, fresh air, which we will examine in a little bit.

Table 7.1 is a simple worksheet to calculate the heat-load within the room. If you put this all together and use the formulas in the table, you'll see that the total Btu's per hour of generated heat are 11,639.7.

Table 7.1. Heat-Load Calculation Chart A

ITEM	REQUIRED INPUT	CALCULATION	HEAT OUTPUT
Gear	Full Power Load in Watts	Total Wattage Load	1493.0 WATTS
Lighting	Room Area in Square Feet	Floor Area x 2.0	640.0 WATTS
Power Distribution	Power System Rating in Watts	(0.02 x The Power System Rating) + (0.02 x Gear Above)	22.8 WATTS
People	Maximum Occupancy Load	160.94 x each person	804.7 WATTS
Total		Total From Above	3410.5 WATTS
Conversion		Multiply Wattage Total x 3.4129	11,639.7 Btu's / Hr.

It takes 12,000 Btu's to create one ton of cooling, so your needs are .97 tons of cooling per hour for the control room.

Now let's add a tracking room to the equation. It's going to be 15x21.5.

Let's start again with a gear list:

- Guitar Amp, 300 W
- Bass Amp, 500 W

- Keyboard Amp, 250 W
- PA System, 500 W (Although you won't use this for recording, if the room doubles as a practice room, you should add it to the equation.)

In a tracking room, everything stays pretty much the same (from a calculation point of view) with the exception of the latent load per person. Musicians can generate as much Btu output as anyone who's working fairly hard, so we will use a met rate of 3.0 for this space (353.56 W per person). Also (in this case), you won't be using an isolation transformer for the space. Let's calculate using a four-piece band working in this room.

If you create another worksheet (Table 7.2), you'll see that the requirements in this room are different from the control room. Although the total load is only about 1/3 greater than the control room, the heat load due to people is fully double. The total load for this room is 1.08 tons.

Table 7.2 Heat-Load Calculation Chart B

ITEM	REQUIRED INPUT	CALCULATION	HEAT OUTPUT
Gear	Full Power Load in Watts	Total Wattage Load	1550.0 WATTS
Lighting	Room Area in Square Feet	Floor Area x 2.0	645.0 WATTS
Power Distribution	Power System Rating in Watts	(0.02 x The Power System Rating) + (0.02 x Gear Above)	0.0 WATTS
People	Maximum Occupancy Load	321.89 x each person	1609.4 WATTS
Total		Total From Above	3804.4 WATTS
Conversion		Multiply Wattage Total x 3.4129	12,984.1 Btu's / Hr.

Remember that you still have to deal with fresh air and what it adds to the equation.

Sensible Loads and Latent Loads

When you deal with cooling requirements, there are two very different loads you need to contend with. One of these you are familiar with, and the other you have probably never encountered.

Sensible loads are what you are used to seeing when you read a thermometer. You want it to be 70°F (21°C), and it's 85°F (29°C). So you need to cool down by 15°F (8°C). Simple right? Yes, it is, it's the actual measurable change in temperature.

Latent loads (on the other hand) are a bit more involved. When dealing with latent loads, what you are actually doing is determining the amount of moisture that needs to be dehumidified, with the understanding that this will not change the temperature in the least.

Latent heat is defined as "heat that when added or removed, causes a change in state, but no change in temperature."

Back to People

For the purpose of air-conditioning design, the total Btu output from people is broken down further into sensible and latent loads. However, this is not linear. It isn't a matter of 60/40% or any other fixed ratio. Just as the greater the activity, the greater the Btu output from your body, the greater percentage of that output is latent heat.

Take a minute and examine Table 7.3. It indicates how all of this goes together for a person. Note that on the right-hand side of the chart, the total output per person is broken down in Btu value as a percentage between sensible and latent output for the various activities we have been discussing.

Table 7.3 Heat-Load Breakdown per Activity

ACTIVITY	MET RATE	WATTAGE OUTPUT (m²)	Conversion to S.F.	WATTAGE OUTPUT (f²)	Btu Output	Watts Out Mr "A"	Conversion to Btu Output	BTU Out Mr. "A"	Sensible Heat%	Latent Heat%	Sensible Btu/h	Latent Btu/h
reclining	0.8	46.56	10.7639104	4.33	14.76	94.2814	3.4129	322	60%	40%	192	130
seated quietly	1	58.2	10.7639104	5.41	18.45	117.852	3.4129	402	66%	34%	267	135
sedentary activity	1.2	69.84	10.7639104	6.49	22.14	141.422	3.4129	483	62%	38%	298	185
standing relaxed	1.2	69.84	10.7639104	6.49	22.14	141.422	3.4129	483	62%	38%	298	185
light activity	1.6	93.12	10.7639104	8.65	29.53	188.563	3.4129	644	52%	48%	334	310
meduim activity	2	139.68	10.7639104	12.98	44.29	282.844	3.4129	965	31%	69%	295	670
high activity	3	174.6	10.7639104	16.22	55.36	353.555	3.4129	1,207	30%	70%	357	850

Fresh Air

A good rule of thumb for providing fresh air is 15cfm (cubic feet of air per minute) per person. So in keeping with our five people, 75cfm would be required for the control room and 60cfm for the sound room. This air is also going to have to be conditioned.

Accurately calculating the amount of humidity that fresh air adds to the "mix" here is no easy feat, but to help you understand the basics, I'll pass on some information from ASHRAE (the American Society of Heating, Refrigerating, and Air-Conditioning Engineers).

In 1997, ASHRAE developed what they referred to as the "Ventilation Load Index" (VLI), which is the total load generated (in one year) by fresh air, calculated at the rate of 1cfm that is supplied to a building from the outside.

This index consists of two numbers, which separate the loads into dehumidification (latent) and cooling (sensible) loads for easy comparison. Thus it reads: latent ton-hours per cfm per year and sensible ton-hours per cfm per year.

Understanding the system is easy. For example, in Hartford, CT, the latent load is 3.0 ton-hours cfm per year and the sensible load is 0.3 ton-hours cfm per year. Based on your control room fresh air supply of 75cfm, that would translate to 225 tons (2,700,000Btu's) of dehumidification required per year and 22.5 tons (270,000Btu's) of cooling required per year.

Likewise, in Key West, Florida, it would be 21.6 + 3.5, indicating a need for 1,620 tons (19,440,000Btu's) of dehumidification required per year and 262.5 tons (3,150,000Btu's) of cooling required per year. One more example, and we'll move on. In Reno, Nevada, it's 0.0 + 0.8, thus 0 tons (0.0Btu's) of dehumidification required per year and only 60 tons (9,600Btu's) of cooling required per year.

The reason you have to pay attention to this is due to the fact that what fresh air adds to the room (unlike gear, people, and lights) is not a constant. As you can see, it can be anywhere from no real concern (Reno) to a huge concern (Key West), or anywhere in between (Hartford).

If you want to check what the index is for your area, you can download the ASHRAE document at the Web site of the Energy Resources Center, University of Illinois at Chicago. Search for "Dehumidification and Cooling Loads From Ventilation Air." Understand one thing, this is not a factor you plug into your design calculation. It is simply a tool you can use to determine the extent of concern you should have with the design side of your system. The designers of the system will have to base their calculations on peak design loads.

Why Should I Care About Humidity?

Once again, looking at those ever-so-average adult human beings, they produce about 2.8kg of vapor per day, which is roughly 98.77oz. So our five people will produce about 14kg or almost 494oz. (3.86 gallons U.S.) of vapor per day.

That's on top of whatever is brought in with the outside air, and all of this in a room that you want to maintain at a maximum of about 45% relative humidity.

In order to understand relative humidity, I want you to picture blowing up a balloon. You already know that you exhale vapor (among other things), so when you blow up that balloon, you will breathe a certain number of grains of vapor into it. The actual number is not important. After you blow the balloon up, look at it (in your mind's eye, of course, after all, you are just picturing this, aren't you?). It's nice and plump because you did a good job of blowing it up. Now tie and seal the end.

Take the balloon and place it in your freezer. Let it sit there for a few hours. When you finally go look at it, you'll find that it looks as if someone let some of the air out. They didn't, but it still looks like it. However, the moisture within the balloon remains constant. The actual grains of vapor within the balloon did not change, but the amount of vapor relative to the volume of the air did change. If the balloon (in the state it was in when you blew it up) was 1c.f. of volume with 400 grains of vapor, and in the cold state was 1/10c.f. of volume, it would still contain the same 400 grains of vapor.

So the amount of vapor relative to the volume increases, hence the term "relative humidity."

Picture now pulling 75cfm of fresh 95°F air with a relative humidity of 85% and placing it into your refrigerator of a room. The relative humidity is going to rise tremendously. You could literally end up with water running down your walls if it isn't properly dehumidified. When air is 50% saturated, it contains only one-half the amount of water that it can contain at the same temperature and pressure. As the relative humidity approaches 100%, the air can take on less and less moisture, and at 100% relative humidity, that air cannot hold more water.

Relative humidity is determined by means of wet-bulb and dry-bulb thermometers. The dry-bulb temperature is the temperature of air as determined by a standard thermometer. The wet-bulb temperature is determined by tying a wet wick over the bulb dipped in a reservoir containing distilled water. Airflow around the wick causes the evaporation of moisture, thus lowering the temperature and producing a reading lower than that on the dry-bulb thermometer.

By taking the differences between these two readings and referring to a psychrometric chart, you can determine the relative humidity in a space. A psychrometric chart is nothing more than a graphical representation of several interrelated air parameters brought together. For a completely thorough explanation of the use of these charts, please visit the following Web site located at the University of Connecticut library (UCONN):

www.sp.uconn.edu/~mdarre/NE-127/NewFiles/psychrometric_inset.html

A sling psychrometer is a tool often used to do this. This is the most common type of hygrometer. Figure 7.2 is a sling psychrometer manufactured by Bacharach, Inc. It's designed to accurately determine the percent of relative humidity without the necessity of consulting complex tables. There is no need to wet the wick each time a reading is taken, and it contains a slide rule calculator, which correlates wet- and dry-bulb thermometer indications for direct reading of relative humidity. When not in use, the thermometer case telescopes into the handle for protection. You open it up and spin it around for a few minutes to get your readings; then adjust the slide rule to see the humidity level.

Figure 7.2
A typical sling psychrometer.

There are also easier ways to do this. Digital psychrometers and hygrometers are available with pricing running anywhere from around $100 to over $1,000. The $100 instruments are only usually accurate to ±5%rh. To get accuracy to ±2%rh, the instruments cost around $300.

Being able to accurately measure the relative humidity in your space makes sense once you get things up and running. This way, you will have the information you need to fine-tune your HVAC systems. Something else you have to be concerned with (when considering humidity levels) is the potential for mold growth.

Mold

Let's try to keep this short. Mold tends to become a problem when relative humidity is around or above 50% for any length in time.

There are more than 100,000 mold species in the world. Molds are microscopic organisms that produce enzymes to digest organic matter and spores in order to reproduce. These organisms are part of the fungi kingdom, a realm shared with mushrooms, yeast, and mildew. In nature, mold plays a key role in the decomposition of leaves, wood, and other plant debris. However, problems arise when mold starts digesting organic materials we don't want them to, such as our homes. Once mold spores settle in your home, they need moisture to begin growing and digesting whatever they are growing on. There are molds that can grow on wood, ceiling tiles, wallpaper, paints, carpet, sheet rock, and insulation. When excess moisture or water builds up in your home from say, a leaky roof, high humidity, or flooding, conditions are often ideal for mold growth. Long-standing moisture or high humidity conditions and mold growth go together. Realistically, there is no way to get rid of all mold and mold spores from your home; the way to control mold growth is to control moisture.

The goal is to keep the rh in your space between 40% and 45%. This is close to ideal for most instruments and will help to keep mold at bay.

Understanding the System as a Whole

How does this actually work?

This is where things get "sticky." When it comes to HVAC design, you have sensible cooling loads and latent cooling loads. Sensible cooling relates directly to the capability of the unit to cool the air, while latent cooling relates to the capability of the unit to dehumidify the air. Gear and lighting loads do not add humidity to the air in a space, but people do, as can fresh outside air.

It's important for you to understand the sensible versus latent loads in your rooms and to purchase equipment that meets those needs, which is no small challenge. It might surprise you to know this, but the vast majority of mechanical contractors get this wrong, even on something as "simple" as a home. They tend to use calculation methods that are questionable at best and then add "fudge factors" to cover themselves.

If their calculations come to $3 \, 1/2$ tons of cooling for your home, they'll tell you to get 4, explaining to you that this will make sure that on even the hottest days of summer you'll always have plenty of cooling. Although this is partly true, it's also very wrong. It's partly true because you will never have to worry about having enough cooling. It's very wrong because it isn't the right kind of cooling. What happens is that the unit is so oversized that it super-cools your home, utilizing very short runtimes, which means that you don't get enough air moved across the coils to dehumidify your home properly. This also could cause additional wear and tear on your compressor motor and higher electrical bills for you.

Air conditioners are generally inefficient in the first stages of operation, while their efficiency increases the longer they run. If the on-time of an air conditioner is only five minutes, the efficiency (EER) is somewhere around 6.2. On the other hand, with a properly sized air conditioner, the same amount of cooling would take place in a little less than 10 minutes, and the efficiency would rise to 6.9. This additional runtime would save the customer about 10% of their energy costs.

The capability of an air conditioner to remove moisture (i.e., its latent capacity) is lowest at the beginning of an air conditioner cycle. In order to remove moisture from the air, the coil has to cool to the point of creating dew point, and then it has to have time to cycle the room through complete air changes. As the warm moist air passes over the coils, the moisture wets the coil and (when the unit runs long enough) flows off the coil into the drip pan. It is then removed, either through the use of gravity drains or with condensate pumps. When the compressor short cycles, the moisture on the coil does not complete condensing, and when the unit stops, the moisture on the coil evaporates back into the indoor air.

Volume versus Velocity

The volume of air you push into a room is measured as cubic feet per minute (cfm), whereas the velocity (the speed of air through the ducts) is measured as feet per minute (fpm). We'll use cfm and fpm for the remainder of this conversation.

One of the tricks to creating a quiet HVAC system is to maintain large amounts of volume with low speed. So, where a typical air-handler system (by "typical," I mean a normal home system or a system in an office) might have velocities in excess of 1,000fpm (850 is preferable), in a studio system I like to see velocities below 300fpm, and if at all possible, 100fpm. This isn't that difficult to obtain and still maintain the same volume required for the room. You just increase both the duct size and the number of supply ducts to provide the volume required.

Picture that in a typical 4" round duct you can achieve 100cfm with velocity of 1,167fpm with static pressure of 3/4" in 100' of pipe. Now, if we lower the velocity to 300fpm and keep everything else relative, a duct size of 13.5" diameter will handle the job. So that's one of the goals of the studio design: high volumes with low velocities.

System Options

There are a wide variety of systems available for cooling your recording studio. Now that you understand the basics of design and the amount of money you plan to devote to your system, you can look at the options you have to get the job done.

Split/Packaged Direct-Expansion (DX) Air Conditioners

Figure 7.3 is a typical setup for a split AC system. With these units, an air-handler is installed inside the building, while the condenser/compressor for the unit sits outside. This helps to keep the sounds from the condensing unit's fan in the compressor from entering your space. It's important to locate the condenser so that it is free from shrubbery and other conditions, as that would reduce the free flow of air to the unit. This would reduce the unit's efficiency.

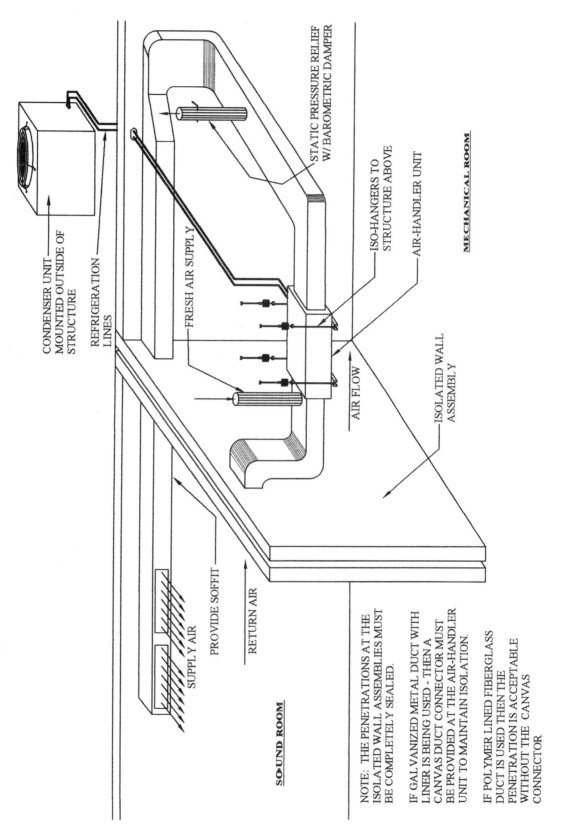

Figure 7.3 Isometric view of a split air-conditioning system.

Note the use of isolation hangers for the air-handler unit inside the building. This stops the transfer of vibration from the system into the structure above. Also, note that the duct work is connected to the system through a series of 90° bends in the duct work, which is also important in helping to reduce the transmission of air-handler noise (fan motor, etc.) into the room.

Fresh air is introduced into the system by a ducted connection (to outside air) immediately adjacent to the unit on the return side of the fan. A barometric release (which is also ducted to outside air) is connected into the system to maintain a small amount of positive pressure in the room, while allowing any pressure buildup beyond that to release. This is an important part of the system. Otherwise, the room will achieve the static pressure capacity of the fan and will no longer be able to bring fresh air into the space.

A supply duct running adjacent to the wall/ceiling assembly helps to maintain isolation from above while maximizing ceiling heights in the room.

As far as the duct material you choose to use, it can be lined galvanized duct or a polymer-lined fiberglass duct board, but understand that the lining is an important part of achieving sound attenuation for the system as a whole. Don't forget, if the duct is galvanized, a canvas connector is required between the duct and air-handler unit to stop the transmission of mechanical vibrations from the fan assembly through the body of the duct work. Fiberglass duct board will not transmit those sounds.

Through-the-Wall Systems

In the tracking room, you can always use a window or through-the-wall unit, if you take care to construct a manner of closing it off when you record. A through-the-wall unit is an air conditioner that installs in a sleeve that goes through a framed opening in the wall. These are similar in nature to the units you see installed in windows, but a through-the-wall unit is a year-round installation, whereas the window unit is typically temporary.

If you picture it, the amount of time it takes you to record is only a few minutes, after which you can always open things back up and turn the unit back on.

This doesn't really lend itself well to a control room situation, however, due to the fact that you will sometimes spend hours on end while mixing down, with only seconds between stopping and starting between songs, or working the mix on the same songs.

Figure 7.4 is one manner of dealing with the penetration and still maintaining isolation. This is also a very effective isolation system for windows in general.

Chapter 7 HVAC Design Concepts

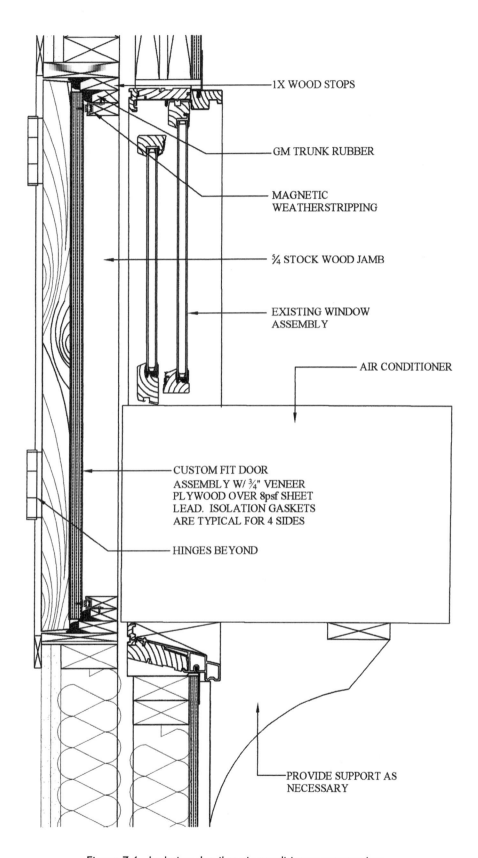

Figure 7.4 Isolation detail at air conditioner penetration.

Ductless Mini-Split Systems

Mini-split systems offer fairly high efficiency and reduced noise without a large hole in the wall or an open window. By separating the compressor and condenser coil from the fan and evaporator coil, the noisiest component is away from the room. The indoor unit will usually have remote control capabilities and a timer to cycle the system only when needed. The indoor unit is still called an air handler because it has the evaporator coil, blower, and controls inside. The outdoor unit is still called the condenser. They are connected together with refrigerant piping and control wiring, similar to a standard split system. Some of these units employ very quiet fans for the air-handler unit and would be quite compatible with the needs of either control or tracking rooms.

Some manufacturers use low voltage to control the system, while still others use line voltage. Caution must be taken when opening up the cabinet to shut power off when servicing. The most important service item is dirt. Screens or filters can usually be found behind the front grille of the air handler. Tabs allow them to slide out for cleaning. As with the condenser unit on split systems, keep vegetation and debris away from the outdoor unit to allow good air flow for maximum cooling efficiency. An occasional blast from a garden hose with the system shut down will help keep the condenser clean.

Relatively new to the American market, ductless split systems have been in use in Japan and other markets for a long time. Until recently, none were of U.S. manufacture, but increased demand has changed that.

If you intend to go this route, make sure to pay attention to your fresh-air needs because these units do not necessarily provide for this. Even the units that claim to provide fresh air must be looked at closely. I have never been able to determine by looking at any manufacturer's literature exactly how much air they are providing, and because you are building airtight rooms, that is an important question. When in doubt, design as if they provided none. What little air they may give you could be fine for a home or office, which has normal air infiltration, lots of occurrences where doors open to allow outside air to refresh, and where windows can be opened. That does not mean there is enough air for your airtight rooms.

Portable Air Conditioners

Portable air conditioners are one of the newest types of air conditioners. Although they still have an exhaust tube that must be vented out somewhere, they are truly portable in the sense that they require no permanent installation.

Portable air conditioners virtually always have caster wheels for portability. They consist of one "box" that holds both the hot and cold side of the air conditioner in one, and they use their exhaust hose (in some cases a duct is required) to expel heat. There are several ways to get rid of the water that the air conditioner condenses out of the air. Most units collect this water in an internal drain bucket, but some exhaust the water through their drain hose while others exhaust the condensate through a duct run to the outside world. Some units have pumps that pump the condensate water out through a drain hose. Most models that collect condensate water in an internal bucket can also be adapted for direct draining.

Since portable air conditioners have both the hot and cold sides of the air-conditioning cycle contained in one box, they have to cool their condenser coil with air they intake from the room they are cooling, and then this air is expelled through an exhaust hose or duct. For this reason, portable air conditioners may create a negative air pressure in the room in which they are run. Air would then be continually seeping into this room from the rest of the house or building to replace the air that the portable air conditioner expels. Some portable air conditioners have solved this problem with a two-hose connection to the outside, one hose to intake air to use to cool the compressor, and one hose to expel this air. In fact, most industrial portable air conditioners use this type of setup. This setup will be required for your space unless you provide another (separate) method for the fresh-air supply.

Some of these systems run quietly enough that operating them in your rooms won't cause any problems. I've heard a lot of home studio owners sing their praises. Your best bet is always to check them out for yourself to decide whether they'll work for you.

Advantages: No ductwork to install, very small penetration to the outside of the room.

Disadvantages: 13,000Btu systems are the largest (at this time) that provide two-hose connections. Fresh air needs must still be met.

Evaporative Coolers

It would be wrong not to cover the evaporative cooler in this book, even though only a very small part of the population can use these systems.

Nature's most efficient means of cooling is through the evaporation of water. Evaporative cooling works on the principle of heat absorption by moisture evaporation. The evaporative cooler draws exterior air into special pads soaked with water, where the air is cooled by evaporation and then circulated. Evaporative cooling is especially well suited to places where the air is hot and humidity is low, and, in effect, acts not only as a means of cooling the air, but also as a means of providing humidity.

Remember that in climates like Reno, Nevada (one of the examples cited previously), the fresh air will add 0% humidity to the room, and your levels will fall well below the 40% to 45% range that you want to maintain in your studio. These units help you achieve that goal. They are available in either portable or ducted arrangements.

One of the nice features of these systems is the fact that there is no compressor, so operating costs are much lower than with standard air-conditioning systems. A disadvantage of these systems is that they require negative pressure in the space in order to work efficiently. An additional exhaust fan will generally be required to provide that condition. This can be interlocked with the unit for operating purposes.

Exchange Chambers

An exchange chamber is a small room adjacent to your studio, which uses a through-the-wall (or window type) air conditioner to pre-condition the air prior to it being exchanged with the air in your space. The air exchange is then handled through the use of exhaust ducts with inline fans to force the air movement you require.

This is a fairly inexpensive method of providing fresh/conditioned air to your space while maintaining isolation. However, it is contingent on your having a wall above grade to install the unit, as well as being able to devote enough space needed to create the chamber itself. Figure 7.5 is one example of how to construct an exchange chamber.

Chapter 7 HVAC Design Concepts

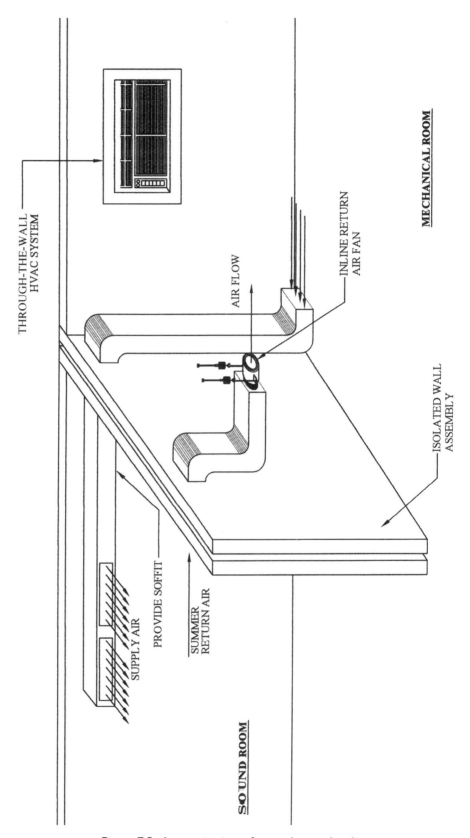

Figure 7.5 Isometric view of an exchange chamber.

147

Combination Cooling/Heating Systems

Until now, we've only been looking at the cooling side of the equation. If you're located in a climate where heating during winter months is a requirement, then it may well make sense for you to consider combination systems that can handle both your cooling and heating needs.

Split Systems

Split system is a term that covers a couple of different type of air-conditioning setups. What it is basically telling you is that there is an interior air-handling unit (basically a fan with a cooling coil) and a separate outside unit. Let's take a look at some of the options you have with these units.

Split Systems with DX Coils

The air-handler unit for a typical split system can be ordered with either an electric or hydrostatic heating coil (in addition to the cooling coil) as a part of the unit. So the only additional parts required for this system would be a connection of additional power for an electric coil or the connection to your boiler for a hot water loop. The existing ductwork is used for circulation purposes.

One thing you should add to your system, if using this for heat, is a low return for the heating season. This is coupled with damper controls to change the air flow between seasons. The reason for this is simple: heat rises. A high return in the cooling season makes perfect sense, peel away the hotter layer near the ceiling, and the cool air drops to the floor.

In the heating season, though, pulling hot air from the ceiling defeats one of the principles of room heat, and that is creating a thermal layer of warmer air until (finally) at 5' above the floor, you reach the temperature the thermostat is set to. In addition, as the air cools, it settles to the floor, and it is this cool air that returns to the coil to be reheated and recirculated through the room. This circulation helps to pull warmer air down toward the floor, making it that much more comfortable in the room. Figure 7.6 is the original system with the return duct modified for winter return.

During the heating season, simply close the damper for the upper register and open the bottom damper. Reverse the process during the cooling season.

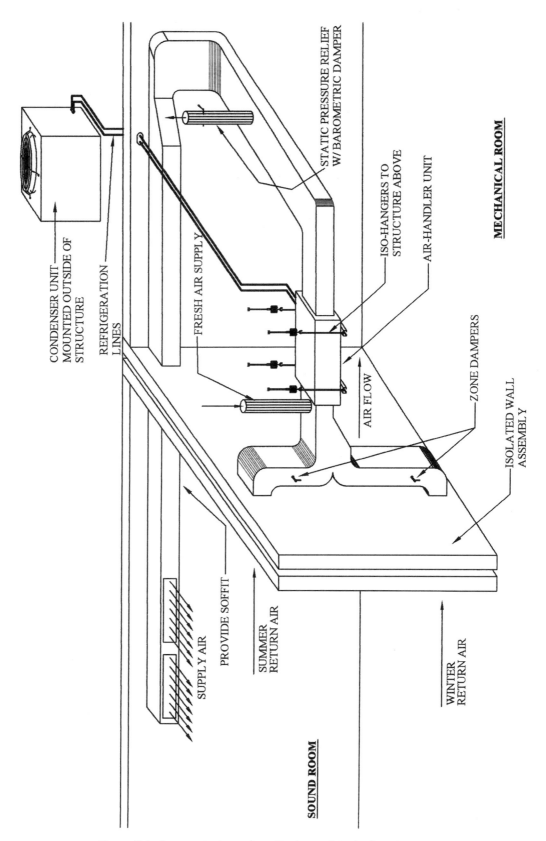

Figure 7.6 Isometric view of a split air-conditioning/heating system.

Split Systems with Heat Pumps

For climates with moderate heating and cooling needs, heat pumps offer an energy-efficient alternative to furnaces and air conditioners. Like your refrigerator, heat pumps use electricity to move heat from a cool space into a warm one, making the cool space cooler and the warm space warmer. During the heating season, heat pumps move heat from the cool outdoors into your warm house; during the cooling season, heat pumps move heat from your cool house into the warm outdoors. Because they move heat rather than generate heat, heat pumps can provide up to four times the amount of energy they consume.

The most common type of heat pump is the air-source heat pump, which transfers heat between your house and the outside air. If you heat with electricity, a heat pump can trim the amount of electricity you use for heating by as much as 30–40%. High-efficiency heat pumps also dehumidify better than standard central air conditioners, resulting in less energy usage and more cooling comfort in summer months. However, the efficiency of most air-source heat pumps as a heat source drops dramatically at lower temperatures, generally making them unsuitable for cold climates, although there are systems that can overcome that problem.

For homes without ducts, air-source heat pumps are also available in a ductless version called a *mini-split heat pump*. In addition, a special type of air-source heat pump, called a *reverse cycle chiller*, generates hot and cold water rather than air, allowing it to be used with radiant floor heating systems in heating mode.

Higher efficiencies are achieved with geothermal (ground-source or water-source) heat pumps, which transfer heat between your house and the ground or a nearby water source. Although they cost more to install, geothermal heat pumps have low operating costs because they take advantage of relatively constant ground or water temperatures. However, the installation depends on the size of your lot, the subsoil, and the landscape. Ground-source or water-source heat pumps can be used in more extreme climatic conditions than air-source heat pumps, and customer satisfaction with the systems is very high.

Through-the-Wall Systems

Through-the-wall HVAC units come in a variety of systems. But they all have one thing in common, they are a complete, self-contained system, and one side of the system is exposed within the room, while the other end is exposed to outside air. A standard air conditioner, the window unit style, is an example of the easiest way to accomplish the installation of one of these units. You can also purchase sleeves that are designed to fit in framed openings in an outside wall for a permanent installation of the equipment.

There are a number of reasons these types of units are preferable to split systems. The first big one is cost. A through-the-wall system can be purchased for roughly 20% of the cost of a similar-sized split system. The next is that they are designed to provide fresh air as a part of their operation, thus handling the need for fresh air within the space. Because of the fact that the fresh air is pulled directly through the cooling coil, dehumidification is handled without the need for additional ducting.

However, one downside is the big hole you just cut through the exterior of your building, and the noise that it will let in and out. Another problem with these units is that the compressor is encompassed within the unit, and, even with quiet units, you can hear these when they turn on.

There are means you can take to use these units unless your actual recording times would be really long. You can install a soundproof enclosure with an interior "door" to close over the opening when you record. You can also tie the power supply into a simple kill switch at your recording desk.

Having looked at the ups and downs to the systems, take a look at the types of systems available to you.

Packaged Terminal Air Conditioner (PTAC)

A PTAC unit is a separate encased combination of heating and cooling that is normally mounted through the wall. It has refrigeration components and forced ventilation, and it utilizes reverse cycle refrigeration as its prime heating source. It usually has another heating supplement, generally an electric element heater for backup in the coldest winter days. This type of unit is usually larger than a typical through-the-wall air conditioner and is often seen in motel rooms and apartment buildings.

Electric Heat Air Conditioners

These units are available in either through-the-wall or window installations. They are the same as standard air conditioner units, with the addition of an electrical heating element for winter months.

Exchange Chambers

Any through-the-wall or window unit can be utilized within these chambers. As with any other ducted system, you want to draw the return air from floor level in the winter and near the ceiling in the summer. One thing is different, though, which is that in the winter you will want to draw supply air from the ceiling of the exchange chamber. Refer to Figure 7.7 for details.

Home Recording Studio: Build It Like the Pros, Second Edition

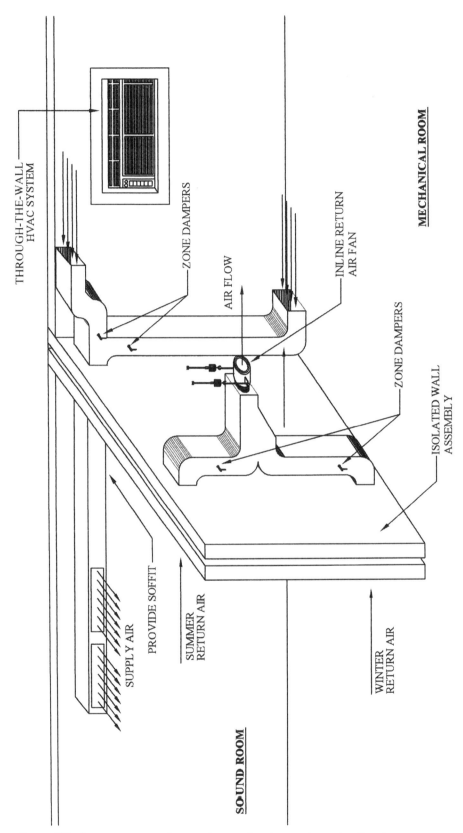

Figure 7.7 Isometric view of an exchange chamber with cooling/heating.

The reason for adding the ability to pull air from the ceiling of the chamber rather than the floor (in winter) is due to simple convection. Heat rises, so the warmest air in the space will be at ceiling level, and this is the air you want to feed into your working room.

Separate Systems

You also have the option of using totally separate systems for your heating and cooling needs. This can be very beneficial if your existing systems can handle the additional loads.

For example, if you have central air conditioning now, and it can handle your needs (which may well be the case, if it is as overdesigned as a lot of home systems are), then adding some duct and a few motorized dampers to handle flow may well be just what the doctor ordered. It is a lot cheaper than a new system by far. You should realize that most home systems do not have the fresh air that you will need for your room, and that you will probably require some additional dehumidification. However, this way is still cheaper in the long run.

The same goes for heat. If your boiler can handle the additional load, it's pretty easy to add some more heating to the space. If you do this, I would recommend that you consider the installation of an under-floor heating system. These systems utilize low-temperature water, using PEX (cross-linked Polyethylene) tubing. They produce heat that is uniform through the room, is much less of a drain on your existing system, and is more energy efficient than standard convection systems (i.e., baseboard heat).

Another benefit to these systems is that they are quiet. Typical hot-water baseboard and electric element heaters go through an expansion process when warming up and can cause quite a bit of noise in your room. Although you could work through this, in the case of a control room, it's possible that mics could pick up the noise in a tracking room and destroy an otherwise perfect take.

With standard convection heating systems, the heating is accomplished by building up thermal layers beginning at the ceiling of the room and working down from there. Eventually, the temperature set at thermostat height (typically, 5' above finish floor) is reached. It isn't unusual to have temperatures 10° to 20° warmer at ceiling height than they are at the thermostat. Also, temperatures at floor level are much cooler than they are 5' in the air.

But with radiant floor heat, the starting temperature at the floor is the controlling factor, with the air cooling off as it rises, and the output from the floor continuing until the thermostat is satisfied.

Not having to develop the thermal layer at the ceiling (before satisfying the needs of the room) means that lower water temperatures can be used, and not having to heat an area of the room that no one comes in contact with (by this we are thinking about the couple of feet directly below the ceiling) means that heating costs are lower. Radiant-floor heating systems typically use water temperatures of 85–140°F (30–60°C), compared with baseboard hydronic systems that operate at 130–160°F (54–71°C) or sometimes 160°–180°F (71–82°C) with older boilers.

Radiant floor systems require special hardware and valve assemblies to work, and they cost more than standard baseboard to install. But if you can afford the initial investment, then the added comfort and slight savings in operating costs would be well worth it.

Oh, one other benefit with these systems, they don't eat up any wall space, so your walls are completely open for placement of furniture, casework, gear, or whatever your heart desires.

System Design

OK, so just how would one go about designing a ducted system for a studio? That's one of the questions that I have been asked repeatedly since the first edition came out, so let's see if we can give you at least the basics.

You know how to choose your equipment. You know your fresh air needs and what you need to do regarding humidity (all that info is provided earlier), so what is needed beyond that?

It's not really rocket science here. In fact it's pretty simple. For a split system, you begin by sizing the outlet from the air handler large enough to reduce the velocity to not more than double your desired total velocity at the outlets. One note of import here: If you run into an obstacle that will not allow the duct to remain full sized for a short distance, you can introduce a reduction of 25% of the duct size (in volume) for a short distance, after which an increase back to the original will return you to full air flow.

Then do the following:

- ▶ Provide lining for all supply and return ducts.
- ▶ Install branch ducting again that is sized to provide no more than double your desired velocity at the outlets.

Or:

- ▶ Provide a baffle system (sound attenuator) just prior to air entering the room that's designed to double the volume of the supply duct (a doubling in volume cuts velocity in half, thus achieving your desired supply register velocity).
- ▶ Use a good-quality supply register to ensure that there is no buffeting of air at the register.

See Figure 7.8 for an example of a plan view for a field-constructed baffle. Figure 7.9 is an isometric view of that same baffle. Note that the outlet is double the size of the inlet, which is necessary in order to maintain the reduced velocity of the air. If they are the same size, then the outlet becomes a bottleneck that reduces the airflow below where you want it.

Figure 7.8
Boxed baffle plan view.

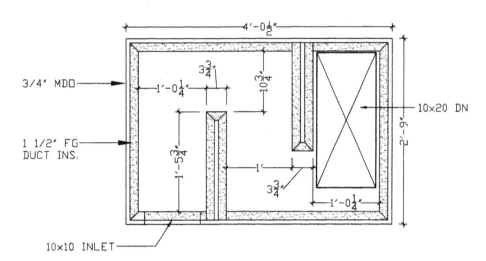

Figure 7.9
Boxed baffle isometric view.

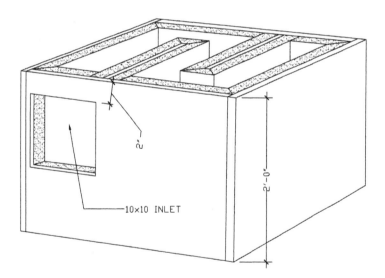

That last part is important unless your velocities are so low that a significant increase in them will not make any difference. The reason is due to the fact that any register introduces a certain restriction in flow. It is not uncommon to see the area of an outlet reduced by as much as 50% with the least expensive Registers/Diffusors. This will effectively double the velocity of your designed outlet.

High-quality registers cost some real money, but in the end they will not introduce noise that you can hear even when they are close to you.

A perfect example of this can be seen in Figure 7.10. This is a high-quality supply register manufactured by Nailor Industries.

Figure 7.10 Nailor Industries Series RBD round diffusor.[3, 5]

The details for this diffusor are shown in Figure 7.11. We'll look at how you can incorporate this in a ceiling cloud in Chapter 10, "Putting It All Together."

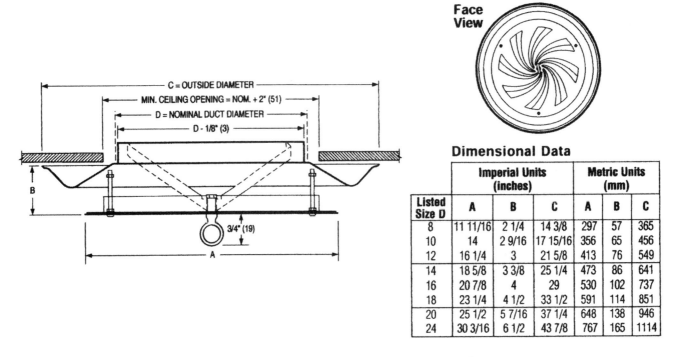

Figure 7.11 Nailor Industries Series RBD details.[3]

Performance data for this product is shown below in Figure 7.12. Note that if you choose properly, you can get some pretty high velocities while maintaining an NC (noise criteria) less than 20, which is barely above the threshold of hearing. For example, with a vertical throw, an 8" diffusor can deliver 314cfm below 20 NC.

Performance Data
Model RDB

Nominal Neck	Neck Velocity, FPM	400	500	600	700	800	900	1000	1200	1400	1600
	Velocity Pressure	0.010	0.016	0.022	0.031	0.040	0.050	0.062	0.090	0.122	0.160
8" Dia.	Airflow, CFM	140	175	209	244	279	314	349	419	489	559
	Total Pressure Horiz.	0.027	0.042	0.062	0.073	0.115	0.140	0.175	0.258	0.335	0.421
	Total Pressure Vert.	0.014	0.024	0.035	0.049	0.053	0.071	0.088	0.122	0.176	0.235
	NC Horiz.	—	—	—	21	23	25	31	33	37	39
	NC Vert.	—	—	—	—	—	—	20	22	28	31
	Throw Horiz.	0-1-2	1-2-4	1-2-5	1-2-7	1-3-9	2-4-10	2-4-11	3-5-12	4-6-13	6-7-15
	Throw Vert.	8	10	16	19	24	31	34	37	43	48
10" Dia.	Airflow, CFM	218	273	327	382	436	491	545	654	764	873
	Total Pressure Horiz.	0.036	0.056	0.082	0.111	0.145	0.185	0.230	0.335	0.462	0.570
	Total Pressure Vert.	0.016	0.026	0.037	0.051	0.066	0.083	0.103	0.149	0.204	0.265
	NC Horiz.	—	—	—	—	21	23	27	33	39	41
	NC Vert.	—	—	—	—	—	—	20	25	32	35
	Throw Horiz.	0-1-3	1-2-5	1-2-7	1-3-8	2-4-10	2-4-11	3-5-12	4-7-13	6-8-15	7-10-16
	Throw Vert.	12	13	22	26	29	34	37	40	48	50
12" Dia.	Airflow, CFM	314	393	471	550	628	707	785	942	1100	1257
	Total Pressure Horiz.	0.047	0.073	0.107	0.149	0.195	0.245	0.307	0.445	0.612	0.800
	Total Pressure Vert.	0.018	0.029	0.042	0.058	0.076	0.095	0.118	0.170	0.232	0.305
	NC Horiz.	—	—	—	33	27	31	35	39	43	46
	NC Vert.	—	—	—	—	—	22	25	28	33	37
	Throw Horiz.	3-6-11	4-7-13	5-8-15	6-10-17	7-11-18	8-12-19	9-13-20	12-16-22	15-18-23	18-20-25
	Throw Vert.	15	17	28	36	46	50	55	60	67	75
14" Dia.	Airflow, CFM	428	535	641	748	855	962	1069	1283	1497	1710
	Total Pressure Horiz.	0.039	0.062	0.090	0.127	0.165	0.209	0.262	0.380	0.542	0.700
	Total Pressure Vert.	0.016	0.027	0.038	0.054	0.070	0.088	0.111	0.162	0.224	0.295
	NC Horiz.	—	—	—	—	22	25	29	37	46	52
	NC Vert.	—	—	—	—	—	—	22	29	35	38
	Throw Horiz.	1-6-12	2-7-14	3-8-16	4-10-17	5-11-18	7-12-19	8-13-20	11-16-22	15-18-23	19-21-25
	Throw Vert.	21	25	31	39	48	53	57	63	70	89
16" Dia.	Airflow, CFM	559	698	838	977	1117	1257	1396	1676	1955	2234
	Total Pressure Horiz.	0.053	0.069	0.110	0.181	0.232	0.292	0.367	0.535	0.737	0.965
	Total Pressure Vert.	0.020	0.032	0.045	0.061	0.083	0.104	0.132	0.189	0.261	0.342
	NC Horiz.	—	—	—	22	25	31	37	42	46	52
	NC Vert.	—	—	—	—	—	22	27	35	39	41
	Throw Horiz.	6-10-18	7-11-20	7-13-21	8-16-22	9-17-24	11-19-25	13-20-26	14-21-27	15-22-28	16-23-29
	Throw Vert.	25	27	34	41	50	55	59	67	85	94
18" Dia.	Airflow, CFM	707	884	1060	1237	1414	1590	1767	2121	2474	2827
	Total Pressure Horiz.	0.071	0.114	0.162	0.226	0.300	0.375	0.472	0.690	0.942	1.230
	Total Pressure Vert.	0.023	0.037	0.053	0.073	0.096	0.120	0.150	0.217	0.298	0.390
	NC Horiz.	—	—	22	34	37	41	44	52	57	62
	NC Vert.	—	—	—	—	—	24	27	33	37	41
	Throw Horiz.	8-13-21	10-14-22	11-16-23	12-17-24	14-18-25	15-19-26	16-20-27	18-22-28	21-23-29	23-25-30
	Throw Vert.	29	34	39	44	55	57	63	74	85	100
20" Dia.	Airflow, CFM	873	1091	1309	1527	1745	1963	2182	2618	3054	3491
	Total Pressure Horiz.	0.074	0.116	0.162	0.221	0.289	0.365	0.442	0.630	0.862	1.120
	Total Pressure Vert.	0.022	0.035	0.050	0.069	0.090	0.115	0.142	0.206	0.284	0.373
	NC Horiz.	—	25	31	34	38	42	45	53	58	62
	NC Vert.	—	—	—	—	23	27	31	36	42	46
	Throw Horiz.	10-14-20	12-16-23	14-19-26	16-21-29	18-23-31	20-25-32	22-27-34	25-30-37	29-34-39	32-37-41
	Throw Vert.	36	42	48	53	58	63	69	81	90	105
24" Dia.	Airflow, CFM	1257	1571	1885	2199	2513	2827	3142	3770	4398	5027
	Total Pressure Horiz.	0.047	0.073	0.104	0.141	0.182	0.229	0.281	0.400	0.540	0.700
	Total Pressure Vert.	0.010	0.016	0.022	0.030	0.040	0.050	0.062	0.090	0.122	0.159
	NC Horiz.	25	30	34	36	42	47	53	62	70	73
	NC Vert.	—	—	—	24	27	33	38	44	47	51
	Throw Horiz.	12-16-22	14-19-26	17-21-30	18-23-32	20-25-33	23-27-36	25-31-37	29-35-40	33-38-42	34-40-47
	Throw Vert.	43	47	50	58	64	69	87	95	99	113

Performance Notes:

1. All pressures are in inches w.g.. To obtain static pressure, subtract the velocity pressure from the total pressure.

2. Horizontal throws are given at 150, 100 and 50 fpm terminal velocities under isothermal conditions with the face fully closed.

3. Vertical throw (projection) is given at 50 fpm terminal velocity under isothermal conditions with the face fully open. For non-isothermal conditions, use the following correction factors:

ΔT Temp. Differential	Correction Factor
20°F Cooling	x 1.40
Isothermal	x 1.00
10°F Heating	x 0.83
20°F Heating	x 0.58
30°F Heating	x 0.53
40°F Heating	x 0.43

4. NC (Noise Criteria) values are based upon 10 dB room absorption, re 10^{-12} watts. Dash (-) in space indicates an NC of less than 20. Values shown are for the horizontal discharge pattern (center closed) and vertical discharge pattern (center fully open).

5. Data derived from tests conducted in accordance with ANSI/ASHRAE Standard 70 – 2006.

Nominal Neck Size	Ak Factors
8	0.13
10	0.25
12	0.51
14	0.56
16	1.08
18	1.36
20	1.60

Figure 7.12 Nailor Industries Series RBD performance data.[3]

Noise-Level Design Guides

In the following chart you can compare the recommended noise criteria levels for various work areas:

RECOMMENDED NC CRITERIA

NC LEVEL	ENVIRONMENT	TYPICAL OCCUPANCY
NC 15 - 20	Extremely quiet environment, supressed speech is noticably audible. These rooms are suitable for acute pickup of all sounds	Broadcasting Studios, Concert Halls, Recording & Movie Studios, Music Rooms.
NC 30	Suitable for very quiet office spacs, large confrence rooms. Telephone use is satisfactory.	Residences, Theaters, Librasries, Executive Offices, Director's Rooms.
NC 35	Quiet offices, satisfactory for confrence room with a 15-foot table, normal voice usage 10 to 30-feet, telephone use is satisfactory.	Private Offices, Schools, Hotel Guestrooms, Courtrooms, Churches, Hospital Rooms.
NC 40	Satisfactory for confrence rooms with a 6 to 8-foot table, normal voice range is 6 to 12-feet, telephone use is satisfactory.	General Office use, Labs, Dining Rooms.
NC 45	Satisfactory for confrence rooms with a 4 to 5-foot table, normal voice is clear at 3 to 6-feet, raised voice at 6 to 12-feet, telephone use is occassionally difficult.	Retail Stores, Cafiterias, Lobby Areas, large Drafting and Engineering Offices, Reception Areas.
> NC 50	unsatisfactory for confrence rooms with more than 2 or 3 persons, normal voice is clear at 1 to 2-feet, raised voices are clear at 3 to 6-feet, telephone use is slightly difficult.	Computer Rooms, Steno Pools, Print Machine Rooms.

My goal in studio design is to keep the noise criteria below NC 20, preferably at or below NC 15 if at all possible, which is just slightly above the threshold of hearing. Take a look at Figure 7.13, which is a comparison of the dB levels of various NC ratings for frequencies ranging from 63 to 4,000Hz.

Figure 7.14 lists the sound-pressure levels for various noise criteria levels from NC 15 to NC 65 in increments of 5.

All in all, you can see that there is plenty you (or your HVAC contractor) can do to make the HVAC system in your studio quiet (be it home or pro). As long as you stick to the basics, you'll do fine. Just remember that high volumes and low velocities are your friends in this endeavor, and don't forget to maintain your room isolation in the process.

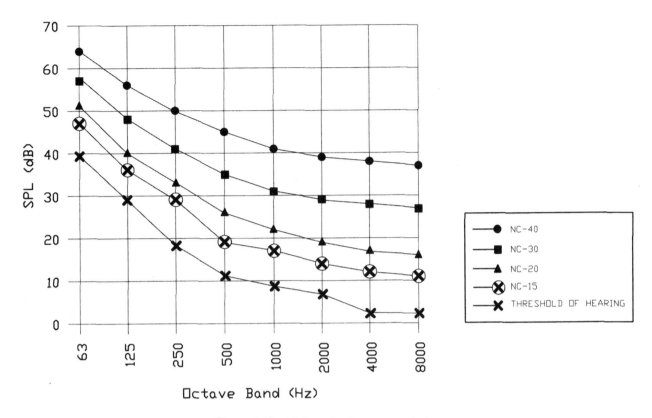

Figure 7.13 Noise criteria curve analysis.

Noise Criterion	Octave Band Center Frequency (Hz)							
	63	125	250	500	1000	2000	4000	8000
	Sound Pressure Levels (dB)							
NC-15	47	36	29	22	17	14	12	11
NC-20	51	40	33	26	22	19	17	16
NC-25	54	44	37	31	27	24	22	21
NC-30	57	48	41	35	31	29	28	27
NC-35	60	52	45	40	36	34	33	32
NC-40	64	56	50	45	41	39	38	37
NC-45	67	60	54	49	46	44	43	42
NC-50	71	64	58	54	51	49	48	47
NC-55	74	67	62	58	56	54	53	52
NC-60	77	71	67	63	61	59	58	57
NC-65	80	75	71	68	66	64	63	62

Figure 7.14 NC sound-pressure levels.

Endnotes

[1] Mosteller RD: "Simplified Calculation of Body Surface Area." *New England Journal of Medicine* Oct 22, 1987, 317(17):1098 (letter).

[2] Lewis G. Harriman III, Dean Plager, Douglas Kosar, ASHRAE Journal, Issue 37, November 1997.

[3] Reprinted with permission from Nailor Industries.

[4] Calculated from William A. Yost and Mead C. Killian, "Hearing Thresholds," *Encyclopedia of Acoustics*, ed. Malcom J. Crocker, (John Wiley & Sons, Inc, 1997), p.1549.

[5] Photography by Dennis Cham, Hit Productions. Reprinted with permission.

CHAPTER 8

Room Testing

In a recording studio, what we hear (the effect that the environment or equipment has on the quality of sound we hear) is everything. What you are going to examine in this chapter is how to use room-measuring software to determine what anomalies exist in your control (listening) room and how effective your treatments are in achieving a good listening environment.

Getting Down the Basics

Let's begin by understanding the reason I say "control (listening) room" and not a recording studio in general.

In a main tracking room, iso-booth, string room, etc., there is not a single specific (targeted) listening area. In those rooms there are multiple locations for microphone placement, depending on exactly what you may be recording.

In tracking rooms, you may be recording anything from amplifiers to vocals. Amps may have microphones placed in close proximity to them, perhaps a combination of microphones intended to pick up direct transmissions and room ambiance.

A vocalist may be a very tall adult or a small child. You may well be recording a saxophone or someone sitting down playing classical guitar. It will often be drums or entire bands.

In all of these cases, there are multiple locations within the rooms that microphones can and will be placed. The microphones in this case are the "ears" in the room. They "listen" and transmit what they "hear" to the recording chain. Although the rooms need to have a good sound, a lot can be accomplished even in fairly bad rooms through the combination of close micing instruments, amps, etc., coupled with gobos to provide at least a semblance of distance between instruments, which will minimize bleed from one instrument track into separate instrument tracks.

A control room, though, is a dedicated listening space. There is typically a fairly small "sweet spot," which is the area that the person handling the engineering side of the recording (which may well include the mixdown of the music recorded) is going to be sitting in. This room needs to be treated in such a manner that the engineer can hear (in the listening position) an accurate representation of whatever is being recorded. Constructing and treating the room in order to obtain this "sweet spot" is the single largest goal in control room construction.

Due to the fact that there are acoustical issues that are psychoacoustic in nature, which means that the mind will interpret certain sounds to be something that they are not, testing a control room is an important part of the whole picture. Room anomalies might be able to trick the mind, but they cannot trick measuring instruments or software.

Sound

Everyone agrees on what good and bad sound quality is. In terms of loudspeakers, this has been proven with double blind listening tests performed by Floyd Toole[1] when he worked in Canada to determine relevant measurements for perceived loudspeaker quality. His work on this spanned a 20-year period beginning in the 1970s.

An important aspect of sound quality (especially in a critical listening environment) is that the equipment reproduces all sounds with the same amplitude at which they were recorded. This requires a frequency response from the equipment (amplifiers, mixing boards, speakers, etc.) capable of providing the same gain for all frequencies. If you have a setup that can achieve this, you should be able to produce a mix that translates in the real world without having to go back and forth between different systems outside of your room to verify the results.

An equally important aspect (to all of this) is the effect that the room has on the sounds transmitted into it. The best gear in the world won't sound good in a room that makes it sound bad.

Subsequent research (by others) indicates that reflections from walls, ceilings, mixing boards, etc. can have effects that can negatively color the sound you hear while you mix, causing you to make poor or incorrect choices in EQ and reverb.

What we are going to examine here is how to test your room to make certain that there are no room anomalies affecting the music you are trying to mix.

The Software

The software we are going to examine here is RPlusD by Acousti Soft, Inc.[2] It is a very reasonably priced package with lots of bells and whistles. This software was developed under the Windows 2000 operating system and works within the XP, Vista, and Windows 7 operating systems.

It has also been run within Windows emulators on Mac machines. However, before purchasing it (to run on a Mac computer), you should download the demo program first (which is the full version, minus the key codes you need to turn on all the features) and test it to make certain it will run using your particular emulator.

There is also a freeware program called Room EQ Wizard that I have heard good things about but have never used. This software was written and tested under Windows XP and then subsequently tested with the Windows Vista operating system. It will also run on the Mac. (It has been tested on both the OS X 10.4 and 10.5 versions of the Macintosh operating systems; however, OS X 10.5 Leopard is the recommended operating system on the Macintosh.)

For the purposes of this book, we will only examine the RPlusD software. It would be much too involved to attempt to examine and compare different software packages, and it would make no sense whatsoever to examine professional stand-alone hardware designed for acoustic testing. Regardless of the software you eventually use, the concepts are the same.

There is an array of standard audio measurements, all of which are provided with RPlusD software. The basic measurement types, along with a brief description of the situation in which they are used, are described in the sections that follow. This chapter will explain the basic operation of audio analyzers, using examples from actual tests performed using RPlusD software.

The use, physical configuration, and external equipment required for use with RPlusD software will be explained. The chapter will conclude with a further explanation of the types of measurements with aspects of experimental procedures required to ensure accurate, precise, and repeatable results.

One great thing about this package is that the manufacturer is only a phone call away, answers his phone, and is right there to help you if you have any issues or questions. He has a Web site you can visit and also has a thread at Ethan Winer's forum where he will directly answer any questions you might have (http://forums.musicplayer.com).

Let's begin with an understanding of what exactly it is you need to measure (and treat) in order to make your room correct, and how you can go about determining this with the software.

Room Anomalies

Room anomalies are all of the various issues you read about in depth in Chapter 2, "Modes, Nodes, and Other Terms of Confusion." These are issues like room modes, flutter echo, early reflections, etc. that will make it difficult, if not impossible, for you to just sit at your desk and mix a song that translates well in the real world.

Early Reflections

An Energy-Time graph is shown in Figure 8.1. This result is used to compare direct sound levels emanating from the loudspeaker with reflection levels. Reflections originate from ceilings, walls, and floors when they are untreated.

The mirror trick is a well-known way of determining offending surfaces. This is done as follows: One person sits in the listening area while a mirror is moved along a hard surface, such as a wall, by a second person. The observer looks at the mirror until the image of the loudspeaker appears. When this happens, the mirror is where the reflection occurs.

Measure from the loudspeaker to the surface at the mirror point. Then measure (from the mirror point) to the listener (the path length) and subtract the loudspeaker-to-listener distance (the direct source). The remaining distance can be converted to a time delay. Sound takes approximately 1ms to travel 1 foot. (If the measurement is in meters, sound takes about 2.91ms to travel a distance of 1 meter.) Multiply the distance difference by the appropriate time to obtain the time delay equivalent. Figure 8.1 is an example of a measurement that indicates direct sound and a subsequent reflection of that same sound. Note that the direct sound from the speaker occurs at t=0 seconds (reflections occur later).

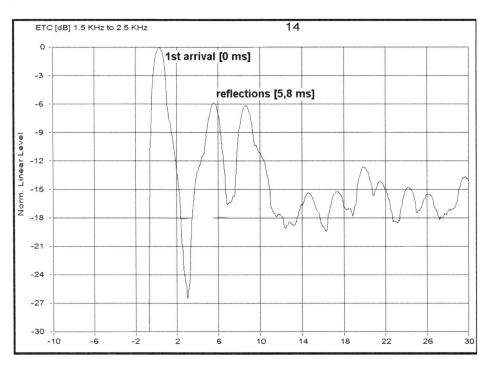

Figure 8.1 Energy-Time graph.

Early reflections (also known as *inter-stimulus delay*s (ISDs) and the *Precedence Effect*[3][4]) are known to create problems with the sound you will hear in a listening room. Early reflections are occurrences where the original signal from your speaker (lead location) combines with a reflected signal (lag location). There are three different phases of this effect:

- ▶ **Summing Localization (Phase I):** This refers to reflections that combine with the original source before about 1ms. These reflections join the original signal with the reflected signal with the effect that a single sound appears to emanate from roughly halfway between the two sound sources. This has the added effect of modifying the original signal with a slight tonal coloring.

- ▶ **Localization Dominance (Phase II):** This occurs roughly from 1ms to 10ms. The signals of the reflected and original sounds combine with the result being a stronger signal that is localized by the first signal source (which would be the monitor in the room, in this case).

- ▶ **Phase III ISDs:** These are reflections that occur after about 10ms, and in this case, two distinct sources of the sound can be perceived, one near the lead and one near the lag.

Notice that the time span between signals is always an approximation. This is due to the fact that although everyone will experience the same results of these reflected effects on the lead sound source, not everyone will experience them in exactly the same time frames. For some people, a Phase I ISD may occur between 0ms and 2ms, etc. However, the above approximations appear (from the testing that has been performed to date) to be reasonably accurate in the majority of cases.

Resonant Sounds

Resonant sounds are the consequence of the shape of the room and how the room dissipates sound energy, and they are the direct cause of peaks and dips in amplitude. Resonant sounds create a ringing effect and can be particularly bothersome, as well as more difficult to combat, than simple early reflections. Fortunately, these can be measured easily.

The degree of coloration due to a resonance is approximately proportional to the area under the curve as illustrated within the results of Figure 8.2, which is a comparison of two resonances that are roughly equal in audibility. You can see that the areas under the two different resonances are about the same, resulting in just about the same degree of coloration heard from each.

Resonances can be masked by other distortions, as well as program source material. Small resonances in the midrange are practically imperceptible when music is being played, but in voices and dialogue they are extremely audible. Resonances in the midrange are very troublesome to those monitoring voice recordings because our hearing is particularly acute when listening to speech. This is in the frequency range of 100Hz–5kHz (known as the "voice range").

In each case, we can hear the distortions, which have the secondary affect of coloring the sound. We may not hear these particular distortions as distortions, and we cannot easily identify them just by listening. Sometimes, one type of distortion can sound very much like another. It is very difficult to determine the sources of coloration without taking measurements.

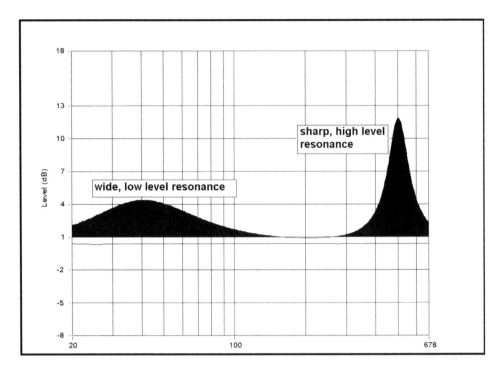

Figure 8.2 Resonance comparison.

Tools of the Trade

To determine the nature of distortions, as well as their likely causes, we can combine a little knowledge about sound with a few measurements and determine the cause. Then, armed with this information and the benefit of experience (or perhaps a good book), we can determine the best or most likely solution to the problem. Then the solution can be tested with another measurement. This is why computer-based audio analyzers are so widely used.

The problem with computer-based audio analyzers is twofold:

- ▶ They are often used incorrectly and give erroneous results because of incorrect setup.
- ▶ The setup and design of the measurement experiment itself is incorrect, leading to irrelevant or misunderstood results.

Sound is something that occurs in both a time realm and a frequency range. The two are intertwined. When we consider the response of a sound system (which in our case includes the room), we do so in terms of time or in terms of frequency. A frequency response and a time response are really the same thing, which is a mathematical representation of how a sound system responds to stimulus. One can be calculated from the other.

Impulse Response

The impulse response is the fundamental measurement from which all other results are determined. It is best explained as follows: Imagine yourself checking the suspension of a car. You might jump on the bumper and watch how the car reacts. If the car bounces up and down numerous times, you know the shocks or springs need replacing. If this data could be recorded, it could be used to determine how the car suspension would respond to any bump in the road using a field of mathematics known as *signal processing*.

Similar to a car whose suspension needs new shocks, in signal processing language, the suspension is said to resonate. The shock absorbers critically damp the resonance by dissipating the energy that is normally cycled between kinetic and potential energy creating the resonance. It is absorbed and converted to heat in the shock absorber. The same action occurs in a sound-absorbing device. These devices are sometimes called *dampers*.

The chart in Figure 8.3 represents the raw data (from a room measurement) that is illustrating the actual response of the system. This result is processed to determine all other types of results, including the ETC curve in the impulse chart. Normally, the impulse response is converted to a frequency response.

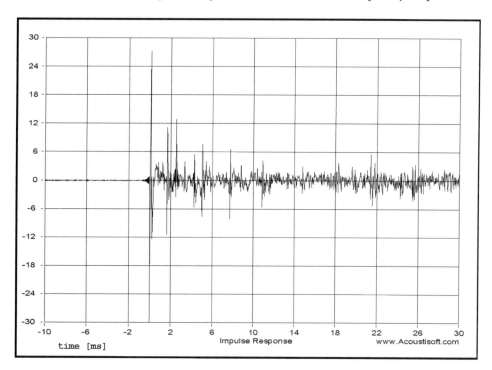

Figure 8.3
Impulse chart.

What is ETC?

The letters ETC are short for "Energy Time Curve," which is a measurement that is representative of the amount of energy released in a given period of time (in this case from the room's response to a sudden burst of impulse energy). Time measurements in milliseconds (ms) are represented on the X-axis of a chart, while the energy levels (dB) are displayed on the Y-axis.

Frequency Response Curve

A frequency response curve is generated by applying a mathematical operation known as the *Fourier Transform* to an impulse response. A Fourier series is an expansion of a periodic function $f(x)$ in terms of an infinite sum of sines and cosines. The Fourier series makes use of the orthogonality relationships of the sine and cosine functions. The computation and study of Fourier series is known as *harmonic analysis* and is extremely useful as a way to break up an arbitrary periodic function into a set of simple terms that can be plugged in, solved individually, and then recombined to obtain a solution to the original problem (or an approximation) to whatever accuracy is desired or practical.

The Fourier series, which is specific to periodic (or finite-domain) functions f(x) (with period 2π), represents these functions as a series of sinusoids, where Fn is the (complex) amplitude. Fourier theorized and later proved that any periodic sound (or non-periodic sound of limited duration) could be represented by Fourier analysis or created out of the sum of a set of pure tones with different frequencies, amplitudes, and phases, known as Fourier synthesis.

Fourier series
$$f(x) = \sum_{n=-\infty}^{\infty} F_n\, e^{inx},$$

Now, having read the previous paragraphs, you're probably thinking that a promise has been broken, that being the promise not to bury you in math. Nope, not true. I am not going to go any deeper into this than I already have. I explained this and gave you a peek at the beginnings of what is a rather lengthy series of mathematical equations just to show you how involved this really is and the lengths you would have to go to in order to work this out on your own. An in-depth study on the theories of acoustical analysis is really beyond the scope of this book, and it would violate the promise I made to you earlier.

I'm giving you just the very basics here, but am happy to say that the program can do all this for you. It can deliver all the information you need without you having to worry about any of the math.

Gating

Gating is sometimes used to erase many of the room reflections from the result before converting it to a frequency response. Figure 8.4 is an example utilizing gating. Notice the high degree of detail in this example. This is sometimes referred to as *grass*, and (as can be seen) when the frequency range increases, the problem increases.

By *problem*, I mean this: Looking at this figure (beginning on the left), you can see that there are a series of very small jagged lines that break off from what could be viewed as the continuity of the main lines in the chart.

These are very small reflections (pieces of *grass*) that serve little purpose in analyzing the data. As you proceed from left to right, you can see that these become more and more obvious and annoying. Around 135Hz, centered from roughly -30dB to -32dB, there are no fewer than 12 of these small reflections. The higher in frequency, the greater the occurrence. By the time you reach around 600Hz, the grass is completely cluttering up the signal.

These small reflections are not only annoying, but they are also (in the long run) completely useless bits of information. They are so small in nature that by the time you finish dealing with the real issues (look at the huge peak and dip around 150Hz for one example), the grass will be gone.

Figure 8.4
Raw frequency response.

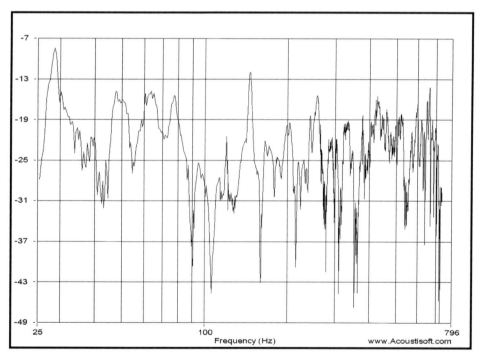

The raw frequency response shown in the figure was directly calculated from an impulse response. It contains far too much detail to be useful. This problem is solved by reducing the applied gate to the impulse before this response is calculated or by using smoothing such as in the fractional octave response. A psychological response analysis uses a combination of these two methods (smoothing and gating) to provide a more accurate indicator of *perceived* spectral balance (if you can't get your head around the perception of sound, please re-read the section on early reflections). However, a psychological response analysis is (computationally) much more intensive than a simple gated or fractional octave response. This really shouldn't matter with the high-speed computers you are working with today. It (the psychological response) washes out detail, such as resonances, which are audible but do not affect the overall subjective frequency balance. Different methods of processing this raw data are useful for different purposes and will be explained in the sections that follow.

Waterfalls

Many people like to use a waterfall response. This response mixes both time and frequency, and it looks very high tech as well as intuitive. The problem with waterfall displays is that they are not as intuitive as they look, and they generate confusion in terms of both time and frequency. They provide a distorted view of reality and are best ignored; however, they are in high demand so makers of analyzers must provide them. A serious user of audio analyzers with understanding would never use or even bother looking at a waterfall plot, except to see if noise is interfering with measurements or perhaps to witness "ringing" of lower frequencies.

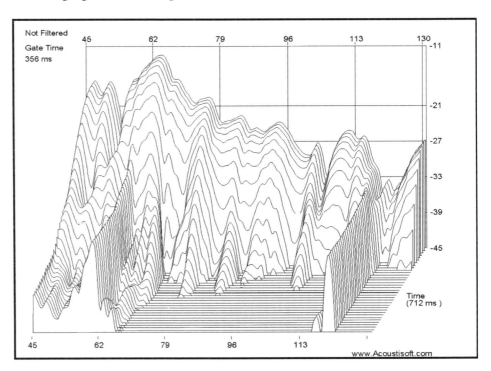

Figure 8.5
Waterfall response 1.

The waterfall response is pretty, but can well be a wolf in sheep's clothing. Changing the parameters changes the display, as can be seen in the figures that follow. There is no single correct set of parameters, making this measurement practically useless in real practice although powerful in visual effect. Compare Figure 8.5 with Figure 8.6, which is the same data reprocessed with different parameters.

Figure 8.6
Waterfall response 2.

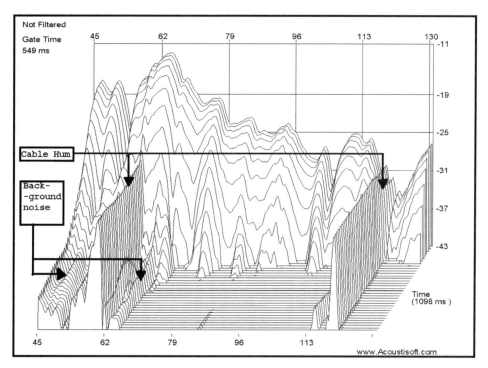

Notice the change in shape of the overall result. Which one is more correct? The answer is neither! Notice also that the extended portions along the time axis in the waterfall display *always* have a frequency response peak at the same frequency that is causing the resonance. The waterfall display dulls the resonance and, depending on settings, obfuscates the true nature of the response. The frequency response works best to quantify the nature of the resonance and is much more easily understood than the waterfall response. Its likely effect on what is actually heard is also much more evident because you can see the area affected by the resonance in the frequency response display, which gives a better indication of its overall audibility as well as a better indication of the exact offending frequency.

Human hearing operates over a very wide range, from approximately 20Hz to nearly 20,000Hz. This is a 1,000:1 difference in frequency, and it is difficult to show on a single graph. Fortunately, when interpreting a frequency response in terms of what is actually heard, this kind of detail is not necessary. The data can be smoothed using a moving average, or in lesser or older products using the less evolved method of forward and reversed IIR filtering of the data, which was the standard before high-speed/high-memory PC-based analyzers became prevalent. Smoothing data results in what is known as a fractional octave smoothed display. Coarse and smooth smoothing are shown in the two figures that follow. Figure 8.7 is an indicator of what to expect with coarse smoothing.

Figure 8.8 is an example of what you should expect to see with a smooth smoothing setting. Notice that the gate time is the same for both results. The reference "should expect to see" relates to the degree of detail displayed and not the actual frequencies. Your physical data will be different because it will be taken from a different room than the sample data.

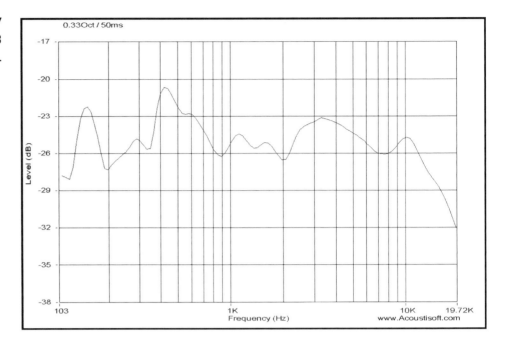

Figure 8.7
Fractional octave display (1/3 octave smoothing).

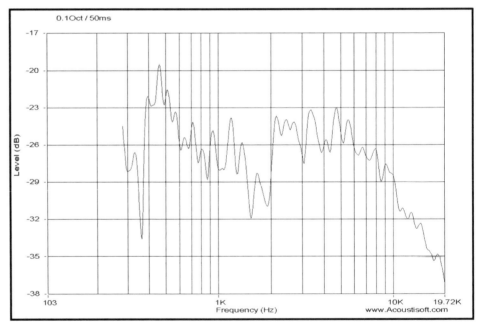

Figure 8.8
Fractional octave display (1/10 octave smoothing).

Notice that 1/10 octave smoothing presents a larger amount of detail than a 1/3 octave smoothing. Also notice that the minimum frequency is much higher while using the same gate time.

A tight resolution such as a 0.1 (1/10) or 0.05 (1/20) octave setting provides the level of detail needed to identify resonances. Whereas a 0.333 (1/3) octave setting is more indicative of overall frequency balance and can be useful for measuring room/speaker power response when enough measurements are taken to provide an accurate average.

Of these two methods used to "clean up" the frequency response, data smoothing (as pictured in the fractional octave results) is used for mid- and high-frequency responses. For room response, more detail is normally sought for the purpose of identifying room resonances. For this reason, you can simply gate the response to exclude the tail end of the impulse response (which eventually becomes just noise) and use the remaining data to calculate a raw unprocessed frequency response. See Figure 8.9 for an example of gating for this purpose.

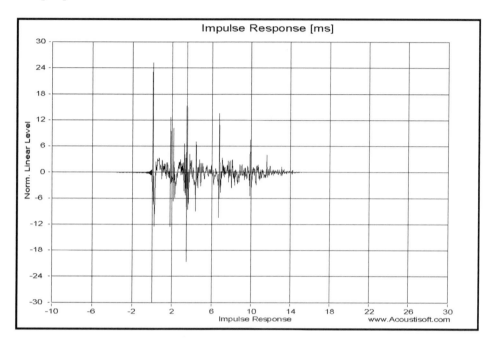

Figure 8.9
Gating the impulse response.

Notice the latter part of the impulse is not shown. This is the action of a gate. Gating (as applied to the impulse response) is meant to illustrate which data is included or excluded with a particular gate time, simply as a "visual aid" to see the nature of the type of gate used in this software (which is more suitable for audio than a conventional Gaussian[5] or Hanning[6] curve). The gate must be adjusted independently for the frequency response when viewing it. The gate here only shows what the gate is actually doing to the data for convenience.

Sometimes, a small gate time is used to exclude all or nearly all room reflections to get a more highly detailed view of the speaker response. This is normally done by knowledgeable users who truly understand the physical nature of systems in terms of mathematical descriptions. Most users would only use the fractional octave result. Fractional octave responses are also useful for fine-tuning and setting equalizers when doing final adjustments to correct for any small remaining acoustic anomalies.

The gate time (shown in Figure 8.10) is only 5.1 milliseconds (ms). Note that the lowest frequency shown in the response is 200Hz. Varying the gate time will change the lowest frequency that can be calculated.

Notice that a 2ms gate time provides resolution down to 500Hz, while a gate time of 500ms provides resolution down to 2Hz. See Figure 8.11 to compare the readings in Figure 8.10 with the results of a longer gate time setting.

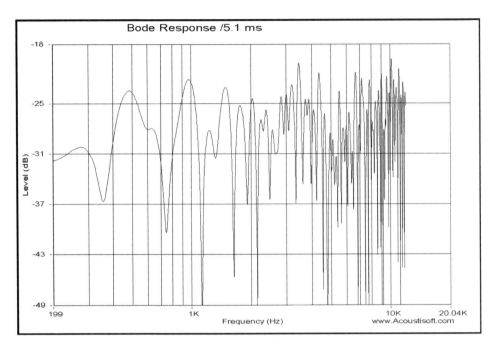

Figure 8.10
Gated frequency response (short gate time).

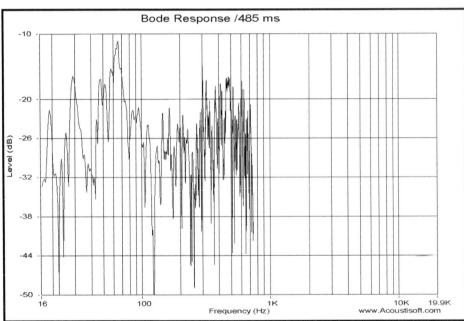

Figure 8.11
Gated frequency response (long gate time).

Longer gate times provide far too much detail at mid and high frequencies to be really useful. Notice the greater degree of detail associated in the longer gate time (485ms in this case). This is complex because a room response itself is highly complex. In fact, this high-frequency data is so complex that it really is useless to display it.

Gate time also affects the lowest frequency that can be shown in the fractional octave results, but a given gate time increases the lowest frequency that can be displayed while resolution is maintained. A 10ms gate time in the fractional octave measurement leads to a lower frequency limit of several hundred Hz rather than 100Hz, as is the case with unsmoothed data (see Figure 8.12). The data displayed is dependent on the resolution chosen while setting the gate time. Longer gate times are necessary with this measurement in order to see results at lower frequencies. Multiple averages tend to smooth out the effect of discrete reflections in the display while still showing their cumulative effect on response. This will be examined in greater depth later in this chapter.

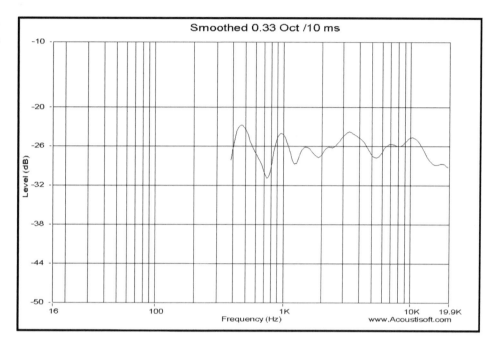

Figure 8.12
10ms gate—smoothed result.

Notice the lowest frequency that can be displayed with the 10ms gate using a resolution of 1/3 octave. Lower frequencies could be displayed, but the resolution would be less than 1/3 octave and therefore results would be misleading.

Psychological Response

The psychological response (shown in Figure 8.13) is significantly more computationally intensive. It varies gate time with frequency (at low frequencies) as required to preserve its 3/4 octave resolution. At frequencies above 200Hz, the gate time is held at 5ms and fractional octave smoothing is applied at higher frequencies to smooth out data. This response is useful when multiple averages are not convenient, and it provides a full range response somewhat more indicative of the frequency balance heard by the listener than could be provided using a setting of 1/3 octave.

Notice that this is a very smooth, full-range response and that the effect of room reflections is removed with the gate time of 5ms at higher frequencies.

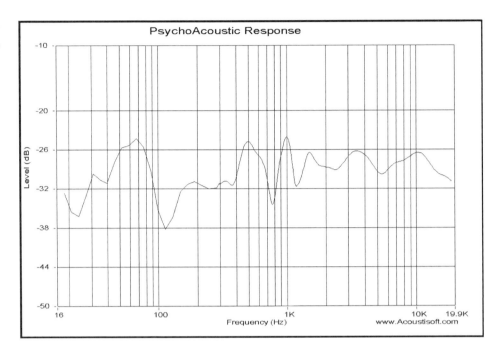

Figure 8.13
Psychological response.

Psychoacoustics.

Psychoacoustics is basically the study of how people perceive sound, and it includes a person's psychological response as well as the physiological impact of sound. In psychoacoustics, the term *sound* would include frequency and vibration. I say "vibration" because certain frequencies can be physically felt as well as heard. A perfect example of this would be how you can feel the bass emanating from the sub-woofers in some cars (turned up to extreme levels) long before the car ever reaches you. Another good example would be how you can feel the bass at some music venues "thump" you in your chest.

Proper Use of the Software

Learning how to set up and use room-testing software properly is critical if you expect the data you gather to be accurate. This is an area where you don't want to rush. If you have gotten to the point where you are ready for room testing, you've already invested a lot of time, energy, and money in your project. Don't throw that away just because it's too much work to read the manuals (or watch the videos) in order to learn how to use the software properly.

The Nature of Scientific Measurements and Experiments

People often rush into using an analyzer, thinking that they can get great measurements at a fraction of the cost. However, they often have little understanding of what the analyzer actually does. Often, big budgets and careful methods are thought to be frivolous and not worth the rates the practitioners charge. Nothing could be farther from the truth. Often, people who are inexperienced in the use of this type of software will end up producing results that only reflect incorrect wiring of the system. Because of this, they end up examining data that is meaningless due to their error. Other results may be generated from some experimental procedure they used that cannot provide measurements that reflect anything even remotely resembling reality, even though the measurements look perfectly valid. How the measurements are obtained is as (or perhaps even more) important than the actual results.

Measurements like those are often imprecise, irrelevant, and not repeatable. In earlier versions of the AcouistiSoft/ETF analyzers, it was not uncommon for Doug Plumb (the software engineer for both ETF and RPlusD, as well as the Owner of AcouistiSoft) to receive emails with questions about certain measurements that were taken by users of that software and recognize immediately that the measurement was meaningless garbage. Yet the user of the software never realized this. Changes to the software in RPlusD will largely prevent this from happening.

A typical measurement setup is shown in Figure 8.14.

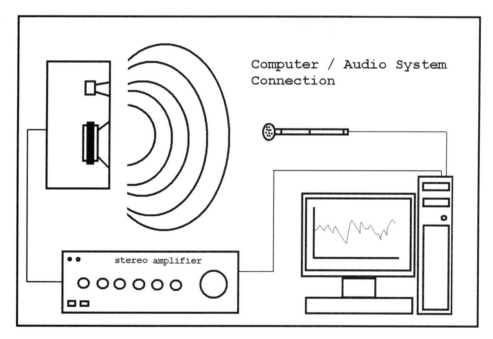

Figure 8.14 Measurement setup.

Notice that the computer software provides an excitation test signal for the audio system input, and that only one channel of information is provided. Only one channel should be measured at any one time because erroneous results occur due to interchannel interference when more than one speaker is playing the test signal during a measurement.

RPlusD is slightly more complex than other analyzers in its setup because it provides numerous checks to ensure that the measured results are relevant and not the result of unrelated phenomenon, such as an MP3 file playing music in the computer, an improperly connected sound card, if the sound card tone controls are not set properly, or even something as simple as a less than ideal sound card.

All of these can produce results that are less than precise, accurate, and repeatable. Care is required to ensure that measurements are both useful and relevant.

RPlusD software provides a system that gives a number of checks to ensure correct results. They are given in the next few sections.

Hardware Connections

The hardware setup to use this system may seem a bit daunting when you first get started, but it really is not as bad as it looks at first glance. If you have any problems getting started, please remember to pay careful attention to the details provided by Acousti-Soft in their various PDF files and the informative support videos they have on their Web site.

Loopback for Left and Right Channels

The loopback test is designed to verify that the sound card is operating correctly and to analyze sound that may be introduced via the sound card and computer system that it is connected to. Figure 8.15 is the connection required for performing a loopback test.

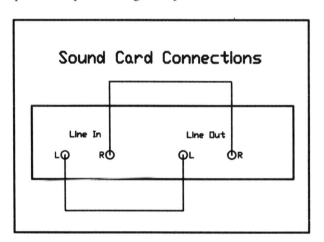

Figure 8.15
Loopback test connections

In addition, it is used to receive mic input and provide a signal to the system. This allows RPlusD to remove any frequency response anomalies that may be caused by the sound card itself or non-flat tone control settings. Tone controls may be set to provide a test signal that boosts certain frequencies that may otherwise be hidden with noise. It does make the software slightly more difficult to set up, but people having trouble with these connections likely are the ones that should have this additional verification. Neophyte users require more checks than those with experience and should play around with the analyzer and get comfortable with this checking of data before taking measurements they intend to rely on.

Chapter 8 Room Testing

Sound Card Connections

The typical wiring diagram for a sound card is shown in Figure 8.16. Many analyzers do not require the left-channel loopback connection, but it is a valuable addition to check the validity of data and sound card calibration, as well as provide the additional benefit of a very accurate mic/speaker distance measurement. (Notice the left-channel loopback connection for sound card calibration and mic-speaker distance referencing.)

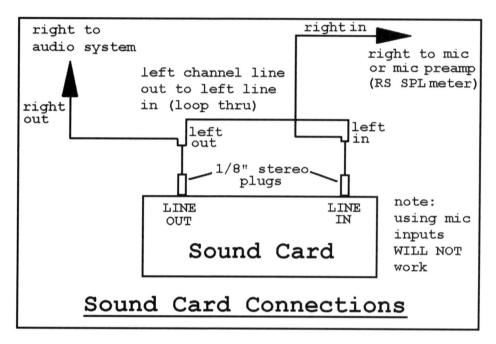

Figure 8.16 Sound card setup.

Data Gathering

Data gathering in RPlusD is a relatively easy process, but there are a few things you need to watch for. This program is relatively well laid out, but you still have to invest of yourself to make certain that the data you gather is accurate. As mentioned previously, this chapter will help you focus on the critical points, but it is ultimately up to you to see to it that this all works the way it is designed to. If you have problems with this, please remember that Acouisti-Soft is only a phone call (or email) away.

The Data Gathering Screen

Figure 8.17 shows the indicators for the feature that estimates the mic-to-speaker distance. This provides a precise mic/speaker distance measurement that can be used to verify results for correctness. If the mic speaker distance isn't correct, then something may be wrong and the measurements may not be correct. Sometimes the impulse needs to be shifted in the software to yield the correct results to get the first peak starting at 0ms. When the impulse is shifted to this position and the mic speaker distance is correct, the shifted measurement can be saved. The error was probably caused by an error in the auto-alignment system in RPlusD, which can never be perfect 100 percent of the time. In some cases, the manual input of mic/speaker distance is required, as in the case of sub-woofer-only measurements that do not provide enough high-frequency information for correct auto alignment to take place.

Sometimes, processing such as that which occurs in an EQ (or another processor) can give an additional delay in the signal, making the mic/speaker distance measure larger than it is in reality. This must be allowed for. It is impossible for a processor to make this distance appear shorter. This feature provides a valuable quick check to ensure the validity of the measurement.

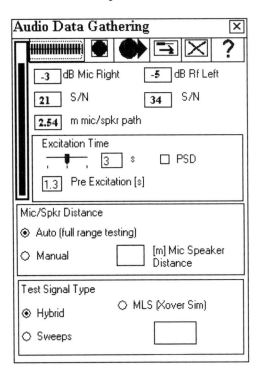

Figure 8.17
Audio data gathering.

Channel Level Indicators

Here are some items you need to pay attention to. In Figure 8.17, you can see the channel level indicators, signal-to-noise estimates, manual/auto mic speaker distance determination, Hybrid/Sweeps signal selection, and the special MLS test signal selection, (which is only used for electronic crossover emulation measurements of loudspeakers).

Pay attention to the values in the various data boxes; these are roughly the numbers that you should expect to see. If you see numbers that vary a great deal from these numbers (for example, positive numbers in the two boxes at the top of the screen), then there is something wrong with your setup.

The MLS test signal selection is only used for electronic crossover emulation measurements of loudspeakers. A discussion of this functionality is beyond the scope of this book and a different subject matter than what you are dealing with in this chapter. For further information on this topic, please visit the AcouistiSoft Web site.

Setup Issues

When I first set this system up for use, there was an issue with my readings during the data gathering process. (The issue was a lot of noise being generated in the final data.) While trying to ferret out the cause of the problem (which included some emails and a few phone calls to Doug at AcouistiSoft), I decided to proceed by methodically disconnecting some outboard gear, more specifically the mic preamp and microphone. (I had no concerns about the outboard soundcard itself as it passed the sound card test with flying colors.) I left the amplifier and speaker hooked up to the sound card so I could check them first. There was no noise with that part of the setup. I then reconnected the mic preamp to verify that condition, again no noise being the result. When I reconnected the microphone to the system the noise was back, and it ended up being as simple as a slightly faulty connection at the point where the mic cable connected to the preamp. Once it was a clean connection, everything worked exactly as indicated in the documentation for the software.

Signal-to-Noise Ratios

The software provides indicators of an estimated signal-to-noise ratio as a piece of the data gathering process. The signal-to-noise ratio shown in Figure 8.17 is normally above 40dB for the loop-through (normally the left channel but this is switchable). Room measurements typically show around 25dB, but in many cases, a 15dB measurement is more than good enough. The test signal need only play at about 80–85dB to obtain these results. It is pointless to play the test signal any louder than what would interfere with normal conversations.

Playing the test signal too loudly may cause hearing damage or other health issues. It is easier to take large numbers of measurements when the test signal is played at a comfortable level. This also avoids user fatigue, which may (in and of itself) be the cause of a large number of erroneous results in data gathering. You can never have too many measurements for an average, but most people take far too few.

Signal Types

There are two different test signal types used with this software, Hybrid and Sweeps, as shown in the data gathering screen in Figure 8.17. Taking two consecutive measurements, one using the Hybrid signal, followed by one using the Sweeps signal is the best way to provide test measurements that will verify that the signals are not being affected by random background noise or distortion.

Comparing these two test signals will reveal the results of parasitic effects that would render the data (otherwise) useless.

Obtaining identical results when comparing these tests is a foolproof way of verifying that the data is unaffected by background noise, electrical noise, distortions in the system being tested, or noise in the room itself.

This preliminary procedure should always be used before running a full array of room measurements. RPlusD measures system gain, so an ideal setup will result in one measurement being directly overlaid with the second measurement, even though the sound-pressure levels of the two test signals will sound different.

It is not a problem adjusting your sound card output levels (if you want to raise or lower the volume during testing) because this will not affect the overall measurement of gain in the system.

During normal room testing, you will generally use the Hybrid signal, as it's much more pleasant to listen to when performing a series of tests.

About Signal-to-Noise Ratios

Figure 8.18 is a frequency graph of an estimate of the signal-to-noise ratio present in the measurement. Conventionally, a signal-to-noise ratio is a rating that describes a piece of equipment rather than a test result. In the case of this software, the estimated noise energy is calculated from the tail end of the impulse and compared with the measured impulse response in the form of a smoothed 1/3 octave frequency response.

This response shows a much larger signal-to-noise ratio at low frequencies. The test signal is designed this way because frequency response at low frequencies tends to show a greater degree of variations and requires a higher signal-to-noise ratio so that test signal energy is much greater than noise. Also notice the erratic curve showing a bad result. (The curves and curve numbers are colored in the actual software package, which makes distinguishing them much easier than what you see here.) It's a crude estimation, but it can make a single bad measurement obvious in a set (as identified in Figure 8.18). This measurement can be deleted before a subsequent average is performed.

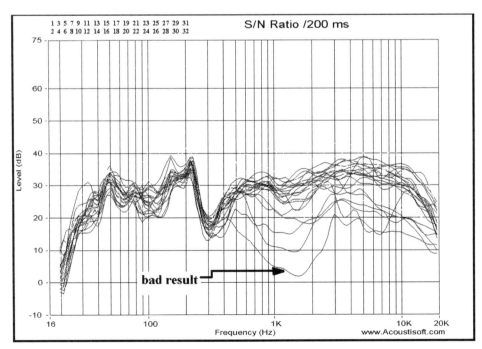

Figure 8.18 SNR data.

SNR—Utilizing the Waterfall Plot

Although I do not find a lot of real value in the waterfall plot (Figure 8.19), it can indicate an approximation of the noise that may be affecting measurements in a similar way as the previous estimate. This is really the only useful result a waterfall plot can provide, and it is no more accurate than the display in Figure 8.18, but it is much prettier. (Notice the regions showing background noise—steady state and random.)

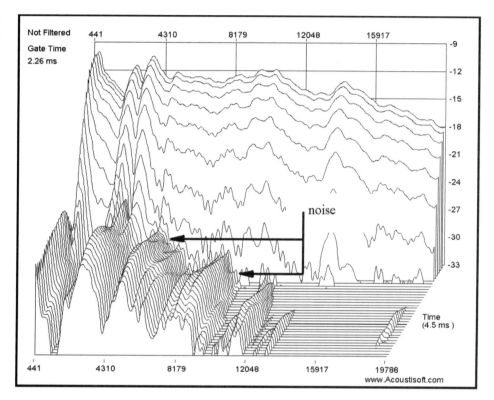

Figure 8.19
Identifying noise with a waterfall plot.

Operation of Analyzers

In order to use this (or any other) room software, you require sound card excitation, which means the connection of the sound card input to a microphone, as well as the connection of the output from the card to some external equipment. Your ultimate goal is (or if it is not it should be) to produce meaningful data that can be reproduced as desired.

Repeatable, Precise, Accurate, and Relevant Measurements

In performing a measurement, the measurement itself must be repeatable and show the same result each time. A person using different equipment and slightly different microphone positions must be able to measure the same thing. The effect of the actual measurement mic location must be filtered out from results if an accurate representation of the system response is to be obtained because listeners never keep their heads in one position when listening to sound systems.

Single microphone measurements may be useful for finding certain system pathologies. For example, a resonance can be measured with a single mic measurement, but the measurement itself really only exposes the particular pathology rather than the response as a whole. Single mic measurements are useful for measuring the magnitude and time delay associated with reflections as well, but this too is a pathology of the system, not its response as a whole.

To measure the room response as a whole, multiple measurements must be taken. This normally consists of placing a mic at one location, taking a measurement, moving the microphone, and then taking another measurement. The correct number of averages must also be used to ensure that there are enough averages to ensure repeatability of the experiment for one who uses slightly different mic locations in the same setting.

Explanations of how to take good measurements are given in the next section with a diagram for each.

Measuring LF Response

A single measurement for low-frequency response can show lots of parasitic effects from room reflections. Two measurements of low-frequency response with different mic positions are shown below. The differences are due to room reflections and these differences are highly dependent on microphone position (see Figures 8.20 and 8.21).

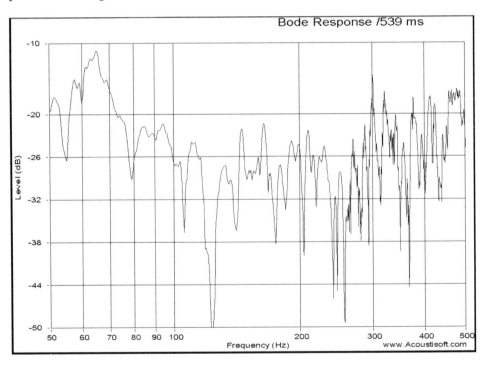

Figure 8.20
Low-frequency response—location 1.

Figure 8.21
Low-frequency response—
location 2.

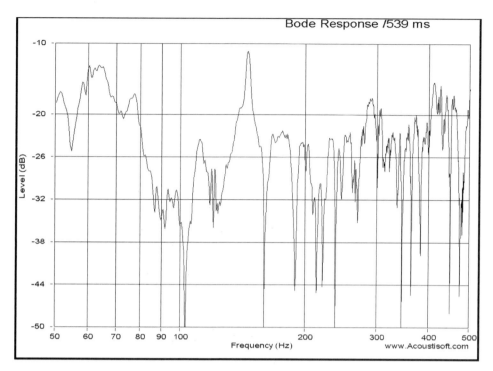

These random parasitic effects can be overcome by taking many measurements and averaging them. This makes resonances more obvious, which is what we want to measure when measuring the low-frequency response of rooms. The low-frequency response of rooms consists of nothing but a set of closely spaced resonances, which are, in fact, the transmission channels in a room (see Figure 8.22). Notice that the room resonances are much "stronger" and more apparent between 40Hz and 70Hz, with smaller issues above 70Hz.

Figure 8.22
Result of averaged LF measurements.

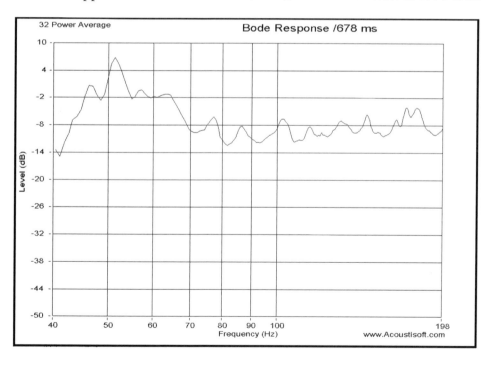

Measuring Your Speaker's Power Response

People often want to measure a room-speaker power response over the full audible range. This requires many measurements and combines the effect of the shape of the room along with the nature of the total loudspeaker response. Many mic locations are used. In this particular test case, there were 32 spread over a 2' wide x 6' long x 3' high area (36 cubic feet).

In a small control room, an appropriate area may be something like a two-foot cube centered at the normal position of the mixing engineer. The spacing between the microphone positions determines the lowest frequency for which this measurement is repeatable. In this series of measurements, the microphone positions were more than one foot apart. The measurement is highly repeatable if the fractional octave smoothing is 1/3 (0.333) octave or lower for frequencies above a minimum wavelength of four times the mic spacing distance. The experimental result below is repeatable for the frequency range above about 250Hz; below 250Hz the results would be slightly different if the experiment were repeated with slightly different mic positions (see Figure 8.23). There is reduced sound power output in the mid-frequency range of this two-way loudspeaker. The speaker actually puts out less total power into the room at midrange frequencies due to the directional nature of the 6.5" woofer producing midrange frequencies. A three-way speaker would perform much better in this regard for that particular frequency region.

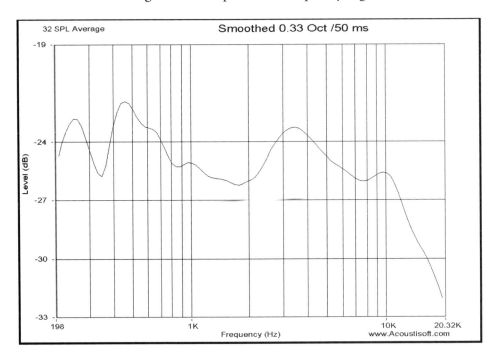

Figure 8.23
SPL room/speaker averaging.

How Many Measurements Are Needed?

A single, or only a few, measurements would indicate a random nature in the data gathered. Due to this, no real conclusions would be able to be drawn. In Figure 8.24, you can see that the individual measurements are all markedly different, but they show a few things in common. Averaging the data filters out any of the random characteristics in the data, leaving us with the commonalities that exist within the various test positions.

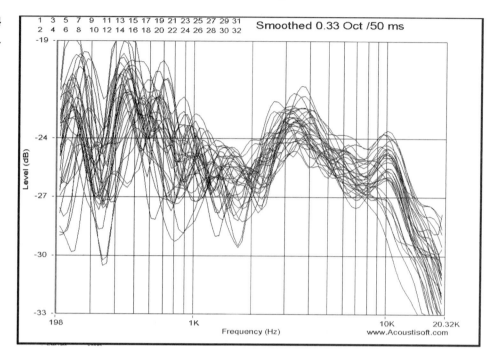

Figure 8.24
SPL data gathering.

If you only want to measure the loudspeaker response, take a lot of measurements (with random positions) near the axis of radiation of the speaker. The response can then be gated to eliminate most of the later room reflections. The few reflections remaining have a different effect for each mic location and will be averaged out. Figure 8.25 is an example of this type of data after averaging using the software.

You can see the smooth response obtained by averaging multiple mic positions. Repeating this experiment with a new set of mic positions within the same area would yield very similar results.

Figure 8.25
Speaker SPL data averaging.

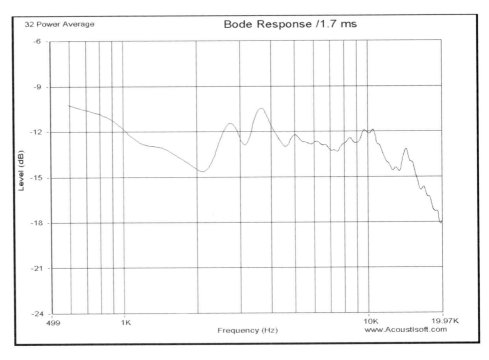

Early Reflections

Early reflections in a control room must be symmetrical and balanced; the RPlusD ETC system gives filtered bands that allow you to see reflected sounds originating from the independent drivers in a loudspeaker system (see Figure 8.26).

Figure 8.26
Early reflections.

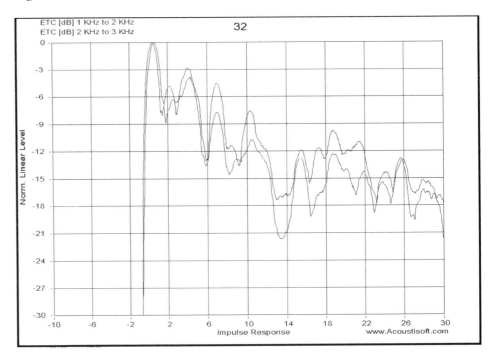

Notice the first reflection from the tweeter is higher than from the midrange. A slight reorientation of the loudspeaker may be all that is required to fix this. This measurement is also useful for testing symmetry in reflections between the left and right channel.

Let's take a moment here and discuss (again) the symmetry of the control room and this particular test.

When performing this test for the purpose of measuring the acoustical symmetry in a room, my personal recommendation would be for you to use a painstakingly methodical approach, using (first) the same speaker and channel for both sides of the room. The reason I say this is because unless you are using some of the best paired speakers around, speakers will often have differences between them. The same is true for the gear you may be using.

Once that signal leaves the computer, even if you split the signal into a mono pair, the entire signal path through your gear and speakers might well produce totally different results left to right. Thus, setting up both speakers (in a symmetrical layout) in a room that is acoustically "perfect" does not necessarily guarantee that testing will indicate room symmetry.

However, for the purposes of this test, if you use the same signal path, meaning the same speaker and system chain, you know that the signal you receive at the microphone is identical side to side (at least from the perspective of the initial signal from the equipment). Then if any anomalies crop up, it is the room that makes the difference, not the equipment. If you are using new speakers and a system that you feel is spot on, then you could well skip this special test. However if something strange seems to crop up, this is always a good test to verify that it is indeed the room that is at fault and not the gear.

Once that is completed, it would make a lot of sense to test the left and right channels fully to verify if the speakers and signal chain are in fact exactly the same. Making minute adjustments then to the EQ of one channel would make it possible for you to fully balance your system. For those of you using state-of-the-art, top-of-the-line gear, this should not be an issue.

Isolation Between Spaces

You probably have gone to great lengths and a lot of expense to try to do this right. From isolated assemblies to room treatments, doing it right is neither inexpensive nor necessarily easy.

Knowing what you achieved in the end (from the perspective of isolation) is probably as important to you as the final room finishes. Testing your room-to-room isolation before installing all of your finishes would make a lot of sense. It's not that difficult to add some layers of drywall before a space is finished, but making a difference after the room is completed would be a huge expense.

Isolation testing is another area where this software shines.

RPlusD is easily used to measure the effectiveness of sound isolation schemes. The procedure involves taking sound-pressure level measurements, with the door to the room being tested in both the closed and open position. This will measure the effectiveness of whatever weather stripping you installed for the door along with the TL values for the wall/door assembly as a whole. Figure 8.27 shows the SPL responses for a typical room taken with the door in both open and closed positions.

Note that in this particular test, it's important to use the manual speaker-mic distance in the software to compare the two, and it is equally important that the dimension entered into the software to identify the microphone location is very accurate.

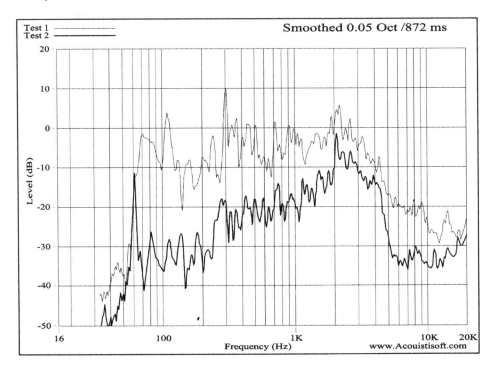

Figure 8.27
Determining room isolation values.

In some cases, wooden doors will act like windows over some narrow frequency ranges, and well-insulated heavy doors will show a uniform decrease in sound level that is perhaps a 20–30dB reduction or even more.

In the particular case above, the test was done using a standard hollow-core door assembly. Test 1 was performed with the door in the open position. Note that the door resonant frequency, as well as the area of coincidence, can clearly be seen in this comparison. The resonant frequency is that spot at just about 62Hz where there is no isolation (the two frequencies join). Coincidence in this door is in the frequency range of 2kHz to roughly 4.5kHz. Notice how little isolation there is in that range. Adding mass, damping, and weatherstripping to this door would definitely help you go a long way toward helping to isolate this space.

You can clearly use this feature to help you assess your success (perhaps further needs) when it comes to isolation.

Parametric EQ

RPlusD also provides a full emulation of the Behringer FBQ 2496 parametric equalizer. The filters in the software duplicate the settings in the equalizer, and the filters are run through the data in the time domain rather than the

EQ curve being added as a delta curve as in other software. RPlusD does not require a measure-EQ-measure-EQ cycle. You measure once, determine your EQ setting in the software, and then apply it to the EQ only once. Precision EQ for midband response is highly beneficial. Problematic room modes are normally best treated with absorbers, but when EQ is the only choice, then RPlusD combined with a Behringer EQ is a very powerful tool, due to the nature of RPlusD and the fine adjustments available of the Behringer unit (see Figure 8.28).

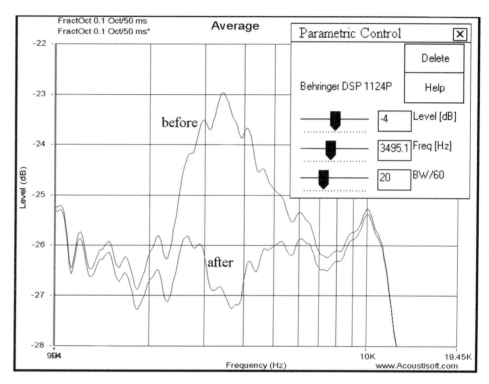

Figure 8.28 EQ settings.

Notice that the software provides the effect of the EQ on any measurement in real time as you adjust the EQ controls in the software while looking at the measurement. RPlusD does not emulate the anti-alias filter in the Behringer unit. Its gentle decline is best compensated for by a single-filter 1dB boost at 20KHz using a 3-octave (180/60) bandwidth setting.

Just a few notes here about parametric EQ and DSP (Digital Signal Processing).

Although the documentation for this software uses a Behringer unit as its example, the crossover filters themselves are universal, meaning that any DSP EQ will behave the same way provided it uses the "Standard Bilinear Transform," which it should for mapping frequencies in and out of the digital domain. This is never mentioned in spec sheets and is always used. There is no reason for a designer of a DSP EQ system to use any other mapping method.

Digital signal processing has come a long way from the time when it started, but it is still not a substitute for a well-designed, well-treated room. However, within a well-treated, well-designed room, it can go a long way toward helping you tweak out that last little bit that makes the difference between an average room and a really good room. (Really great rooms can stand on their own merit.)

So please don't let yourself get sloppy because you think that when all is said and done you can deal with anything remaining by simply using DSP—it just isn't going to happen.

Endnotes

[1] Toole, Floyd E., "Listening Tests: Turning Opinion into Fact," *Journal of the Audio Engineering Society*, National Research Council, Ottawa, Ont. K1A OR6, Canada Vol. 30. Issue 6, pp. 431–445, June 1982; "Loudspeaker Measurements and Their Relationship to Listener Preferences," *Journal of the Audio Engineering Society*, National Research Council, Ottawa, Ont. K1A OR6, Canada, Vol. 34, Issue 4, April 1986 and Vol. 34 Issue 5 pp. 323–348; May 1986.

[2] When the first edition of this book was published, although Acousti Soft, Inc. was in the process of changing their product from EFT to RPlusD, it was not completed by the time of printing—I therefore covered their EFT software in this chapter. The R+D package now replaces ETF, provides the same features as EFT, plus adds some bells and whistles EFT didn't have.

[3] Haas, H., "The influence of a single echo on the audibility of speech," *Journal of the Audio Engineering Society*, Vol. 20, pp. 146–159, March 1972.

[4] "Modeling Physiological and Psychophysical Responses to Precedence Effect Stimuli," 2009, Jing Xia (Department of Cognitive and Neural Systems, Boston University), Andrew Brughera (Department of Biomedical Engineering, Boston University), H. Steven Colburn (Department of Biomedical Engineering, Boston University), Barbra Shinn-Cunningham Ph.D. (Department of Cognitive and Neural Systems, Department of Biomedical Engineering, Boston University).

[5] Named for Carl Friedrich Gauss (*Carolus Fridericus Gauss*) April 30, 1777–February 23, 1855), who was a German mathematician and scientist who contributed significantly to many fields, including number theory, statistics, analysis, differential geometry, geodesy, geophysics, electrostatics, astronomy and optics.

[6] Julius Ferdinand von Hann (March 23, 1839–October 1, 1921) was an Austrian meteorologist. He is seen as the father of modern meteorology. He studied mathematics, chemistry, and physics at the University of Vienna. He was the director of the central institute for meteorology in Vienna (1887–1897), professor of meteorology at the University of Graz (1897–1890) and professor of cosmic physics at the University of Vienna (1890–1910). In signal processing, the Hann window is a window function,

called the Hann function, named after him by R. B. Blackman and John Tukey in "Particular Pairs of Windows," published in "The Measurement of Power Spectra, From the Point of View of Communications Engineering," New York: Dover, 1959, pp. 98–99. In the aforementioned article, the use of the Hann window is called "hanning," e.g., "hanning" a signal is to apply the Hann window to it.

CHAPTER 9

Room Treatments

In this chapter, you'll examine "do it yourself" (DIY) room treatments, as well as learn about some of the options for manufactured units. (These suggestions are from companies that I have dealt with and where I know the people who make things work.)

If you're working your way through this book while you're constructing your studio, then it's probably been a while since you read Chapter 2, "Modes, Nodes, and Other Terms of Confusion." It might not be a bad idea to browse briefly through that chapter as a refresher before reading this one.

First, to make a room sound the best it can possibly sound for the purposes of recording, mixing, and critical listening, acoustical treatments are a necessity. The three key needs are low-frequency control, control of early reflections, and control of sound decay. Thanks to the explosive growth of the personal studio industry over the last few decades, many different types of effective acoustical treatments in many different decorative styles are now commercially available. From DIY to design-build, not only the sound of a room, but also the look and feel—the "vibe"—can be customized.

Low-Frequency Control

"Bass traps," or (more technically correct) "low-frequency control devices," are one of the most important considerations for small-room acoustics. Without good low-frequency control, mixes might sound "muddy" (too much bass) or "thin" (not enough bass—often described as "suck out"). Recordings could have weird resonance "bumps" or cancellation effects (sometimes described as "hollowness").

Effective low-frequency control can smooth things out. After being treated with effective bass traps, the bass that is heard is the true bass from the recording or mix. In mixing, the guesswork is minimized, and mixes translate to other systems much better.

Here are some basic guidelines for choosing various low-frequency devices. The first thing that should be considered is the effective frequency range of the loudspeakers or other sound sources in a room. Placing loudspeakers that roll off below 60Hz in a 12x10x8 room means that controlling the first axial mode at 47.1Hz will probably not be necessary. (Remember that calculating room modes is discussed in Chapter 2.) Placing a bass amp in the same room and recording the performance will require addressing that first mode, since a bass guitar can easily produce frequencies lower than 30Hz if it's of the five- or six-string variety.

After the range of control has been determined, choosing, installing, or building bass traps can begin. Table 9.1 gives some very general guidelines for choosing appropriate low-frequency control for a room.

Table 9.1 Room treatment options[1].

Lower Limit	Typical Room Volume	Example Treatments
60-80Hz	<1,000 ft³	"Standard" corner foam devices Panels across the corner; 1" thick (min.), 12" wide (min.) Small "Helmholtz" resonators
40-60Hz	1,000 to 2,000 ft³	Large corner foam devices Panels across the corner 2" thick (min.), 24" wide (min.) Porous, resonant, or hybrid devices near corner; e.g., 12" diam. tubular "traps" Free-standing porous or hybrid devices around the room Small- to medium-sized "Helmholtz" resonators Porous absorbers (min. 4" thick) on walls and/or ceiling with airspace behind (min. 2" deep)
<40Hz	>2,000 ft³	Larger corner foam devices spaced out from corner (min. 6") "Tuned" panel or membrane devices directly on walls and/or ceiling Deep porous devices (min. 8" thick) on walls and/or ceiling with airspace behind (min. 4" deep) Large "Helmholtz" resonators Passive, electronic damping not in-line with audio signal (i.e., not "room-correction")

Most good commercial bass traps are designed to address a wide range of the low-frequency spectrum. The only limitations tend to be space and money. The latter is less of an issue since there are a lot of DIY approaches that can cost next to nothing. Space, however, can be a limiting factor. Low frequencies have long wavelengths and higher energy in a room (relative to higher frequencies). This usually means that the best low-frequency control devices are going to be large. Since these devices are typically being considered for small rooms, we have a classic "Catch-22" situation. The best low-frequency control could take up more space in a room than all of the gear (and people) combined.

But don't lose hope. Clever designers, manufacturers, and acousticians have been working for years on methods to control low-frequency sounds that don't take up acres of precious studio space. Commercial devices typically take on one of three forms: pressure devices, velocity devices, and hybrid devices that employ a combination of pressure and velocity control. Let's take a look at these.

Pressure Devices

Pressure devices tend to be placed directly in the room corners or directly on the walls and ceiling of a room. The corners of a room are always the areas of high pressure. For certain axial modes, the areas of high-pressure buildup extend to include the entire surface, including the corner. Placing devices designed to control the pressure buildup directly in or near corners is a very common approach to low-frequency control. Devices designed to control velocity buildup can also be used near the corners and surfaces of a room, but should not necessarily be placed directly in the corner or on the wall.

Pressure control devices are the more active types of devices—resonating devices, such as solid panel absorbers, or Helmholtz devices, such as perforated panel absorbers and "slat" absorbers. These devices typically control a fixed range of low-frequency sound dictated by design parameters, such as the size of perforations, the depth of porous cavities, the surface weight of a panel, and so on. In general, a solid or perforated panel, or slats, placed over an empty airspace will be an extremely efficient narrowband absorber. The absorption will be very high at and around the center frequency, but will not affect more than about 1/12–1/6 of an octave around the design frequency. If a porous material is used in the cavity behind the solid panel, perforated panel, or slats, the overall absorption will decrease, but the bandwidth tends to cover about an octave centered at the design (resonant) frequency of the device.

In general, resonant absorbers can be tricky to implement. Care must be taken that certain higher frequencies aren't reradiated into the room by a resonating panel or cavity. It would be a good idea to have professional assistance if you intend to deal with low-frequency issues in this manner, which means that you should hire an acoustician to help with the design and (perhaps) a carpenter to help with the construction.

Velocity Devices

Velocity devices can be as simple as thick, porous materials placed on the wall with airspace behind them. The thicker the material and the deeper the airspace, in general, the more control you will have at lower frequencies. These devices tend to be quite efficient, absorbing sound at and above the lowest frequency predicted, and usually offering some benefits at even lower frequencies. An example of a good velocity absorber is a 4"-thick piece of mineral fiber, typically around 3pcf density, over an equally sized airspace.

Hybrid Devices

There are also quite a variety of devices that work as both pressure and velocity absorbers. These are the more common corner devices offered from many manufacturers. The behavior of a dense, porous material in or over a corner tends to yield benefits both in the pressure and velocity domains. Dense foam absorbers, designed to be placed in corners, can yield control down to 50–100Hz, depending on the size. Spacing the devices away from the corner can offer even lower control.

Dense fiber panels placed over a corner, with or without porous backfill, are also becoming more common. For this you want a wide porous panel placed over a corner (the density of the porous panel used for this should be 6pcf or more) with an air cavity behind it. Filling the space behind the rigid panel with less dense fiber or foam improves the performance even more.

A good example of this type of treatment would be the placement of a 6pcf rigid fiberglass panel around 4 to 6" thick and 2' wide. You want to install this so it straddles the corner by placing it at a 45° angle. Filling behind this with some fluffy insulation would be an added benefit.

Early Reflection Control

To control early reflections, absorption is generally the preferred method. Since the frequency range of these reflections is higher—typically 300–500 Hz and up—thinner materials can be used, which means many more aesthetic choices for room finishes. Absorption for recording studios is typically one of the three "Fs:" fiber, foam, or fabric.

- ▶ **Fiber:** These are usually 3–8pcf mineral (including glass) fibers, as well as other natural (for example, wood, cotton) or synthetic (for example, polyester) fiberboards. The mineral fiber panels are often covered with an acoustically transparent cloth for aesthetic purposes. Most acoustical ceiling tiles also fall into this category, although very few common ceiling tiles are absorptive enough to consider for recording studios. If a ceiling tile grid is going to be included in a recording studio design, careful attention should be paid to the absorption of the proposed tiles in each frequency band (i.e., don't just compare "NRCs"), including the low frequencies. There are many varieties of ceiling tiles available from companies such as Armstrong and USG, some of which have very good absorption characteristics. Ceiling tiles glued directly to a flat wall or ceiling should generally be avoided. The minimum thickness for a flat, fibrous absorber for recordings studios is 1".

- ▶ **Foam:** These are usually 1.5–2.5pcf acoustical foam panels (avoid "packing" or "bedding" foams). There are a wide variety of flat and sculpted designs available for aesthetic purposes. The minimum overall thickness for a sculpted foam absorber for recording studios is 2".

- ▶ **Fabric:** Curtains are the most common in this genre. Heavy drapes (typically greater than 18 oz/yd) with greater than 100% "fullness" might be worth considering. ("Fullness" defines the folds in the drape; 100% fullness means that a drape that is 20' when fully extended with no folds is covering only 10' of wall.) This type of absorber —a moving blanket would be another example—is more effective with an airspace behind it, and it is most effective in multiple layers with airspaces between. The overall thickness of the fabrics plus the airspace should be at least 3" for recording studios, although this approach is typically much less effective than fiber and foam equivalents.

For all of the above, thicker is generally better, at least up to about 4 to 6". If more low-frequency absorption is a desired byproduct of early reflection control, absorbers as thick as 12" would be appropriate. If space on the walls has already been taken up with resonant absorbers for low-frequency control, foam or fiber panels can usually be attached to the faces of them to control high-frequency early reflections. Care should be taken not to add too much mass to a resonant absorber because that could affect the frequency range at which it's most effective. Perforations or slats should generally not be covered up either for the same reason. If you decide to take this approach, then experiment with it for a bit before making the attachments permanent. If you damp the face panel too much, you will reduce its ability to effectively vibrate at its center frequency and thus defeat its ability to perform.

Care should also be taken to choose the absorber that is going to offer the best possible reduction of the early reflection. For the smaller control rooms, angles of incidence between loudspeaker, wall, and listener can be quite high. For flat absorbers, the maximum absorption is observed when sound strikes the panel perpendicular to it, usually referred to as "normal" incidence. A sound source and listener in a small room are usually positioned so that the incident angle of early reflection is not even close to perpendicular. For a flat absorber, the absorption decreases when sound strikes it off-axis (see Figure 9.1). This means that flat absorbers—particularly fibrous absorbers with a 3–8pdf density and a stretched fabric wrap—can actually reflect high-frequency sound. Lower density fibers and foams, particularly sculpted foams, do not exhibit this variation in absorption for angles of incidence. In fact, some newer absorbers have been optimized so that a maximum thickness of material can be positioned to "face" the incident sound and maximize broadband absorption.

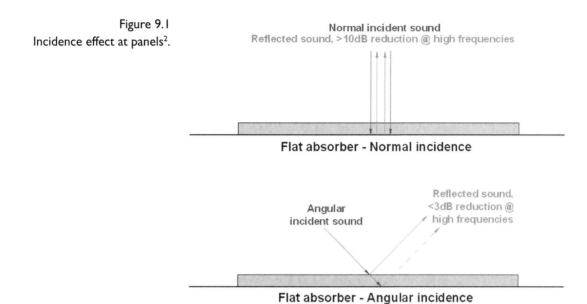

Figure 9.1
Incidence effect at panels[2].

Next, you need to have a basic understanding of what it is you are looking at when you review the data from manufacturers of products you will use to construct traps or manufacturers who provide finished products for your use.

Absorption Coefficient

Understanding the absorption coefficient of the treatments you are going to install in your room is critical for you to treat that space effectively. It is not as difficult as it may seem.

Reported absorption coefficients are based on an area of the material that measures one square foot in area. A reported absorption of 0.00 means the material did not absorb any of the related frequency. The frequency was either reflected or passed through (and back) with no change in levels. An easy way to picture this is that an absorption coefficient of 1.0 represents a 1' square window (in the open position) through the wall of the room, allowing all of that particular frequency to escape the room; however, for some frequencies, the window closes partly, perhaps even completely.

What is even more interesting is that it is possible for the value to be greater than 1. Yup, you can actually capture more sound than 100%. Although this might sound impossible, it's not impossible, and it is true. ASTM C423 ("Standard Test Method for Sound Absorption and Sound Absorption Coefficients by the Reverberation Room Method") does not tell us how to interpret the data when this occurs; in fact, it's completely silent on the subject. So the numbers are just reported as measured.

The generally agreed-upon cause of absorption coefficients that exceed unity (the value 1.0) is effects from diffraction. This means a change in the direction of sound energy in the area of a boundary discontinuity, such as the edge of a reflective or absorptive surface. Thus, if you have an absorber, the front will absorb, and the edges will diffract.

If (at a particular frequency) that absorption measured 10 sabins, with a product that measured 2x4 in size, the reported absorption coefficient would be 1.25 (10 sabins/8s.f. of absorber = 1.25). Thus, if you're looking at products that only have absorption coefficients listed, you can multiply the area of the samples by the absorption coefficient to determine the total sabins for that particular product.

Knowing how to do this (in either direction) will aid you in comparing products from different manufacturers.

Further, any material tested that did not have the edges banded (which is required in ASTM C423) to stop the edge from adding absorption may well exhibit slightly higher values than would be expected otherwise. Make certain to check the test reports of all products to verify that the specimen's were tested in accordance with the appropriate standards. The following description is from ASTM.[3]

4.2 Measurement of a Sound-Absorption Coefficient

"The absorption of the reverberation room is measured as outlined in 4.1 both before and after placing a specimen of material to be tested in the room. The increase in absorption divided by the area of the test specimen is the dimensionless sound absorption coefficient. In inch-pound units, it is reported with the dimensionless "unit" sabin per square foot, Sab/ft^2."

OK, so exactly what is this "dimensionless sabin"?

A sabin is a unit of acoustic absorption equivalent to the absorption by one square foot of a surface that absorbs all incident sound. The unit honors Wallace Clement Ware Sabine (1868–1919), an American physicist and Harvard University professor who founded the systematic study of acoustics about 1895. Sabine used this unit, which he called the open window unit (owu), as early as 1911.

That's where that picture of a one-foot window (mentioned earlier) in the open position comes from—the particular frequency in question would just escape the room. Totally escape = totally absorbed. Pay attention to the fact that I keep repeating "the particular frequency." Remember that it's a special window, because when you open it fully, only some frequencies are able to escape fully, while still others might not escape at all, and others might escape to only varying degrees.

Also, you need to understand that the reported numbers are not percentages. The sabin is a representative value, not an absolute.

One other point that is very important for you to get a handle on is that all of this data is intended to give you a tool you can use to make an intelligent decision as to what product you might want to purchase, based on performance versus cost. Appearance may well be another factor you want to consider, although appearance should probably take a back seat to performance in the long run.

You should not expect to get the same results in your room that some lab was able to obtain under rigid test conditions. They have a very strict set of rules that pertain to the construction of the rooms the tests were performed in, as well as the equipment being used, and for physical conditions within the space (temperature, humidity) at the time the tests were performed.

Those last two are equally important for testing purposes because absorption rates can be greatly influenced by the absorption properties that the air itself has. The bodies that govern the standards are very concerned that these tests be repeatable across the board in any lab certified to perform the tests.

I would like to point out that even with all of the strict control requirements that form the standards, there can still be some huge variations. Cox & D'Antonio reported as much as ±0.1–0.2 for one sample as measured in 24 different labs! (See Figure 3.4 of the first edition of *Acoustic Absorbers and Diffusers*.)

So use the tool as intended, not with the thought that this is what you can achieve, but rather, just how product "A" compares with product "B."

Now, let's take a look at some treatment methods you might use.

DIY Treatments

One important thing to consider with DIY treatments is that the products you use have been tested and certified as being safe (from a flame perspective) in the manner that you will use it.

For example, ASTM E-84 is a fire test standard for construction materials, including wall coverings. This test standard is used to certify fabric, vinyl, or other applied wall coverings that are applied directly to a wall surface.

That is an important part of that particular standard; the materials must be installed touching the surface of the wall. This standard does not certify wall coverings as meeting the standard if they are to be installed with an airspace between them and the surface that they are covering.

The test standard used for that condition (examples would be draperies, table cloths, etc.) is NFPA 701.

So for panel traps, which are constructed with the fabric tightly installed to the fiberglass, ASTM E-84 would be an acceptable standard, while NFPA 701 would not. Likewise, for the mid- and high-bass traps, Helmholtz Resonators, etc., NFPA 701 is used, and ASTM E-84 is unacceptable. Some products are treated, tested, and approved for both standards. This is the best of all possible worlds when it comes to your safety.

The same goes with foams. Acoustic foams are products that have been tested and approved for use under ASTM E-84 (which is similar to NPA #255 and UL #723). These products do not develop smoke or have a flame-spread index (how quickly the flame spreads on the material) that overly endangers life and limb. For example, Red Oak flooring has a flame-spread index of 100, with a developed smoke index of 100. Auralex 2" Studio Foam has a flame-spread index of 35, and a developed smoke index of 350.

Maximum smoke index allowed for interior finish use is 450.

However, standard packing foams (like the foam placed for acoustic treatment that was involved in some of the nightclub fires in recent years) flash and burn like wildfire, with developed smoke well in excess of 450. So keep it safe and only use materials that have been certified to conform to these standards.

Now that this has been said, understand that DIY treatments can be anything from extremely easy to painstakingly exasperating to construct. Some of the easiest treatments you can use are through the use of rigid fiberglass panels. So let's begin by looking at those.

Fiberglass Panels

Products such as Owens Corning 703 and 705 rigid fiberglass panels can be used to create some very effective broadband attenuators. These products (and products manufactured by other manufacturers with the same, or very close, absorption values) can be wrapped in fabric to create not only effective, but also attractive attenuators.

Table 9.2 is an absorption chart for the 703 and 705 products manufactured by Owens Corning. Although the test data only lists frequencies as low as 125Hz, you can see the low-frequency trends with this material as the density (thickness) and mounting changes.

As you can see from this chart, you generally have a slight advantage when using 703 (3pcf density), compared to the 705 (6pcf density), pretty much across the board. Note also that the 6" of 703 product using a Type "A" mounting exhibits absorption coefficients greater than 1.00 at all reported frequencies—a perfect example of edge effect (via diffraction) in play.

Although you are looking at Owens Corning products as examples here, there are a multitude of manufacturers around the world that produce rigid fiberglass panels, and the test data tends to prove them to be fairly equal. As long as you stick with the same densities as the OC products, you will always be in the ballpark.

Table 9.2 Owens Corning Absorption Coefficients

OWENS CORNING 703/705 Rigid Fiberglass									
PRODUCT	THICKNESS	MOUNTING	DENSITY	1/3 Octave Band Center Frequency (Hz)						
				125 Hz	250 Hz	500 Hz	1000 Hz	2000 Hz	4000 Hz	NRC
703, plain	1" (25mm)	Type "A"	3.0 pcf (48 kg/m3)	0.11	0.28	0.68	0.9	0.93	0.96	0.7
705, plain	1" (25mm)	Type "A"	6.0 pcf (96 kg/m3)	0.02	0.27	0.63	0.85	0.93	0.95	0.65
703, plain	2" (51mm)	Type "A"	3.0 pcf (48 kg/m3)	0.17	0.86	1.14	1.07	1.02	0.98	1
705, plain	2" (51mm)	Type "A"	6.0 pcf (96 kg/m3)	0.16	0.71	1.02	1.01	0.99	0.99	0.95
703, plain	3" (76mm)	Type "A"	3.0 pcf (48 kg/m3)	0.53	1.19	1.21	1.08	1.01	1.04	1.1
705, plain	3" (76mm)	Type "A"	6.0 pcf (96 kg/m3)	0.54	1.12	1.23	1.07	1.01	1.05	1.1
703, plain	4" (102mm)	Type "A"	3.0 pcf (48 kg/m3)	0.84	1.24	1.24	1.08	1	0.97	1.15
705, plain	4" (102mm)	Type "A"	6.0 pcf (96 kg/m3)	0.75	1.19	1.17	1.05	0.97	0.98	1.1
703, plain	6" (152mm)	Type "A"	3.0 pcf (48 kg/m3)	1.19	1.21	1.13	1.05	1.04	1.04	1.1
703, FRK	1" (25mm)	Type "A"	3.0 pcf (48 kg/m3)	0.18	0.75	0.58	0.72	0.62	0.35	0.65
705, FRK	1" (25mm)	Type "A"	6.0 pcf (96 kg/m3)	0.27	0.66	0.33	0.66	0.51	0.41	0.55
703, FRK	2" (51mm)	Type "A"	3.0 pcf (48 kg/m3)	0.63	0.56	0.95	0.79	0.6	0.35	0.75
705, FRK	2" (51mm)	Type "A"	6.0 pcf (96 kg/m3)	0.6	0.5	0.63	0.82	0.45	0.34	0.6
703, FRK	3" (76mm)	Type "A"	3.0 pcf (48 kg/m3)	0.84	0.88	0.86	0.71	0.52	0.26	0.75
705, FRK	3" (76mm)	Type "A"	6.0 pcf (96 kg/m3)	0.66	0.46	0.47	0.4	0.52	0.31	0.45
703, FRK	4" (102mm)	Type "A"	3.0 pcf (48 kg/m3)	0.88	0.9	0.84	0.71	0.49	0.23	0.75
705, FRK	4" (102mm)	Type "A"	6.0 pcf (96 kg/m3)	0.65	0.52	0.42	0.36	0.49	0.31	0.45
703, ASJ	1" (25mm)	Type "A"	3.0 pcf (48 kg/m3)	0.17	0.71	0.59	0.68	0.54	0.3	0.65
705, ASJ	1" (25mm)	Type "A"	6.0 pcf (96 kg/m3)	0.2	0.64	0.33	0.56	0.54	0.33	0.5
703, ASJ	2" (51mm)	Type "A"	3.0 pcf (48 kg/m3)	0.47	0.62	1.01	0.81	0.51	0.32	0.75
705, ASJ	2" (51mm)	Type "A"	6.0 pcf (96 kg/m3)	0.58	0.49	0.73	0.76	0.55	0.35	0.65
703, plain	1" (25mm)	Type E-405	3.0 pcf (48 kg/m3)	0.65	0.94	0.76	0.98	1	1.14	0.9
705, plain	1" (25mm)	Type E-405	6.0 pcf (96 kg/m3)	0.68	0.91	0.78	0.97	1.05	1.18	0.95
703, plain	2" (51mm)	Type E-405	3.0 pcf (48 kg/m3)	0.66	0.95	1.06	1.11	1.09	1.18	1.05
705, plain	2" (51mm)	Type E-405	6.0 pcf (96 kg/m3)	0.62	0.95	0.98	1.07	1.09	1.22	1
703, plain	3" (76mm)	Type E-405	3.0 pcf (48 kg/m3)	0.66	0.93	1.13	1.1	1.11	1.14	1.05
705, plain	3" (76mm)	Type E-405	6.0 pcf (96 kg/m3)	0.66	0.92	1.11	1.12	1.1	1.19	1.05
703, plain	4" (102mm)	Type E-405	3.0 pcf (48 kg/m3)	0.65	1.01	1.2	1.14	1.1	1.16	1.1
705, plain	4" (102mm)	Type E-405	6.0 pcf (96 kg/m3)	0.59	0.91	1.15	1.11	1.11	1.19	1.1

0.00 = no absorption.
1.00 = 100% absorption.
Type "A" Mounting - Material is placed directly against a solid backing.
Type E-405 Mounting - Material is placed over a 16" air space. Facing (if noted) is exposed to sound source.

As a general rule, don't get too hung up on small variances between manufacturers because a difference of 15% is virtually meaningless in mid- and high-frequency ranges. However, in the low-frequency range, which is more difficult to handle, it would probably make sense to opt for a product with an absorption rate of .92 instead of one tested at .80. In the end, though, this is a judgment call you have to make based on the amount of low-frequency treatments you need in your room coupled with the cost of the products involved and your general budget.

If you have problems obtaining rigid fiberglass in your area, you can substitute "Fire-Safing" (also known as Rockwool and Thermafiber) for this purpose. Note that Fire-Safing is more prone to breaking up than rigid fiberglass, and it will require a frame for longevity's sake.

Figure 9.2 is a simple 1"x 4" wood frame constructed for a 2'x4' panel using 2" rigid fiberglass. Note the corner blocking installed to keep this square. The corner blocking is simply 3/4" pine let into the routed back of the frame. The use of this material helps you maintain a square frame for your fiberglass. It also gives you a place to mount the stand-off brackets, which will keep this away from your wall, giving you greater sound attenuation. Note the line struck two inches from the outside face and the nails tacked in along that line. With rigid insulation, you can use this method to hold the fiberglass panel in place.

Figure 9.2
Fiberglass panel frame.

Figure 9.3 is the frame with insulation installed.

Figure 9.4 is the back fabric installed. Be careful not to pull too tightly when installing the fabric, or you'll bend the frame. You want the material snug without creases, but not too tight. Experiment with it, and you'll get used to it in no time.

The face fabric is then installed. It's just wrapped and stapled to the back of the panel frame, as shown in Figure 9.5. Take your time with the folds to keep them clean and neat.

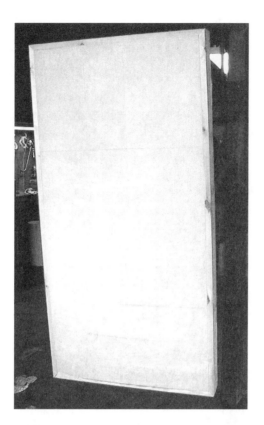

Figure 9.3
Fiberglass panel—frame with insulation.

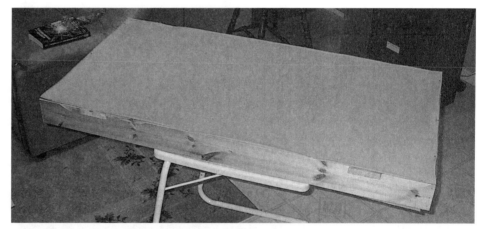

Figure 9.4
Fiberglass panel backing installed.

Figure 9.5
Fiberglass panel facing.

In Figure 9.6, the finished product has simple hangers and 2" standoffs to maintain an airspace between the wall and the back of the panel when installed. (Remember that you gain in efficiency the greater the airspace is behind the panel.)

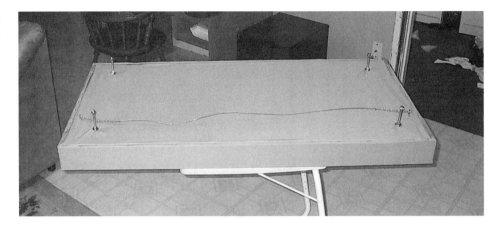

Figure 9.6
Panel standoffs and hangers.

Figure 9.7 is the finished panel in place. Pretty nice looking for a homegrown remedy. Also, in this installation, there is actually a thermostat mounted on the wall behind the panel. The thermostat was installed off-center of the wall and always looked out of place. The panel hides it effectively, but the clearance off the wall allows the free flow of air, so it will still operate properly.

Figure 9.7
Completed fiberglass panel.

Something you have to consider, for this type of panel, is the cloth covering that you use.

In order for these panels to be effective, air has to be able to pass freely through them. This means that you need to have a fabric that "breathes easily." In other words, if you put the fabric tight against your face, then you should be able to breathe easily through the fabric.

One thing of importance—do not go to your local fabric store and pick up just any old fabric for this. Once again, make certain that the fabric you purchase is fire retardant; otherwise, you have to treat it yourself prior to its use. Remember what you learned previously about testing and rating standards.

An excellent source for fire retardant fabrics is Guilford of Maine Fabric (see note below).

Features of Guilford of Maine Fabric

Material is available by the linear yard in a 66" width.

48 standard colors to choose from.

Other Guilford of Maine fabric charts are available for special orders.

Fabric is 100% Terratex polyester.

Pattern is non-directional.

PRODUCT DATA

Flammability: Class 1 Fire Rated per ASTM E-84

Weight: .5oz per linear yard

Tensile Strength: 150 lb. minimum

Tear Strength: 30 lb. minimum

Colorfastness to Light: 40 hours

Colorfastness to Crocking: Class 4 min.—dry; Class 4 min.— wet

Moisture Regain— .5% maximum

Acoustical Solutions, Inc.

Ph: 800-782-5742 Fax: 804-346-8808

Web site: http://www.acousticalsolutions.com

Note that the above product is fire rated per ASTM E-84, which makes it acceptable for this use.

If you choose to use a material that has not been certified by the manufacturer, then you must purchase products made specifically to treat the fabric so that it will be fire retardant. There are a lot of companies that manufacture these products, one of which is Flame Stop, Inc. Figure 9.8 lists their applicable products, and the materials they can treat.

Figure 9.8
Flame Stop, Inc. products.

Flame Stop, Inc.		
MATERIALS	TESTING STANDARDS	PRODUCT
Draperies, Wallpaper, Carpet, Furniture, Decorative Woods, Paper, Straw, Most Synthetics, Acoustical Ceilings, Indoor	ASTM-3 84, NFPA 255, 701 UL 723	FLAME STOP I ™
Fabrics with at least 25% Natural Fiber, Drapes, Bedding, Carpets and Wall Coverings, Silk	TITLE 19, UL 723, ASTM E-84, NFPA 255, 701, FAR 25.853	FLAME STOP I-C ™
Open Cell Foams, Airline Fabrics, Foam Rubber	TITLE 4, FAA 25.853-C	FLAME STOP I-D ™
Open and Closed Cell Foams, Synthetic Fabrics, Plastics, Hay Barriers, Thatching, Bamboo, Tough to Penetrate Materials (Our Most Concentrated Product)	ASTM E-84, NFPA 255, 701, UL 723, FAR 25.853	FLAME STOP I-DS ™

Flame Stop, Inc.
924 Bluemound Rd.
Fort Worth, TX 76131
Tel. 817-306-1222
Fax. 817-306-1733
Toll Free: 1-877-397-7867
Website: http://www.flamestop.com/

Low-Bass Panel Traps

A panel absorber is manufactured by constructing a sealed compartment with a rigid panel face. Typically, your existing wall will create the back of this bass trap, although you can create movable panels as well utilizing a ³/₄" plywood or MDF back surface.

One advantage to a panel absorber is that they can be "tuned" fairly closely to specific frequencies. If after treating your space with broadband sound attenuation, you find that you have one problem frequency "hanging around," then you can deal with that through the use of this type of treatment.

This trap's strength is also its weakness because it cannot be used as a wide broadband treatment. Therefore, it should be used as a last resort and not a starting point.

Figure 9.9 shows plan details for construction of a 2'x4' panel trap using ¹/₄" plywood.

Figure 9.9
Panel trap plan.

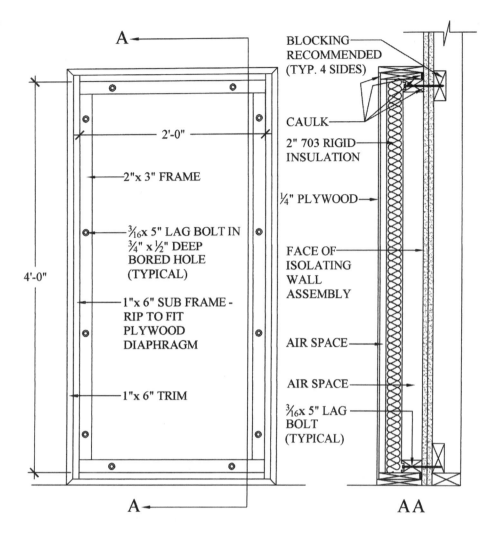

The factors determining the trap's center frequency are the mass (density) of the panel and the depth of the trap.

The formula for calculating the trap frequency is shown in Figure 9.10

Figure 9.10
Panel trap formula.

$$f = \sqrt{\frac{170}{m*d}}$$

Where
f = reasonance frequency
m = surface density of the panel in lbs./sq. ft.
d = the depth of the airspace in inches.

Note that there's no deep math here. This is an easy calculation you can program into a spreadsheet to examine different material properties and panel depths.

Figure 9.11 is a chart of typical plywood sizes and their corresponding densities. Note that these numbers are per design, but you should really accurately measure the panels you purchase due to the fact that different core material and facings, as well as different manufacturers, can provide materials of the same thicknesses with different densities.

Figure 9.11
Typical plywood properties.

Typical Plywood Densities						
3.3 mm	=	4.85 kg/sheet	1/8 =	10.70 Lbs/sheet	=	0.33 psf
6.5 mm	=	9.71 kg/sheet	1/4 in. =	21.40 Lbs/sheet	=	0.67 psf
7.5 mm	=	11.20 kg/sheet	5/16 in. =	24.69 Lbs/sheet	=	0.77 psf
9.5 mm	=	14.10 kg/sheet	3/8 in. =	31.09 Lbs/sheet	=	0.97 psf
12.5 mm	=	18.60 kg/sheet	1/2 in. =	41.01 Lbs/sheet	=	1.28 psf
15.5 mm	=	23.10 kg/sheet	5/8 in. =	50.93 Lbs/sheet	=	1.59 psf
18.5 mm	=	27.50 kg/sheet	3/4 in. =	60.63 Lbs/sheet	=	1.89 psf
25.5 mm	=	38.00 kg/sheet	1 in. =	83.78 Lbs/sheet	=	2.62 psf
Information from the Canadian Plywood Association.						

The panel shown previously in Figure 9.9 has a center frequency of roughly 90.64Hz, based on the densities calculated in Figure 9.11. If you were to deepen the trap by 2" (substitute 1x8 for the 1x6 frame), the center frequency would drop to 77.13Hz. Likewise, if you were to leave the frame as is and substitute 1/8" plywood for the 1/4" plywood, the center frequency would raise to 129.16Hz. Also, understand that the placement of the insulation behind the panel gives you attenuation that covers about one octave around the center frequency (half above–half below). If you do not install the insulation, the panel frequency tightens to roughly one-half octave (one-quarter above and below).

Be careful when you build one of these to maintain the airspace between the panel and fiberglass as shown. If the fiberglass is allowed to touch the back of the panel, it will damp the panel, and the trap will not perform as designed. It is equally important that this trap be 100% sealed against air movement. The trap works on an air-spring principle: any air leakage will cause the trap to fail. So caulk in all locations indicated thoroughly. Allow caulk to cure for a minimum of 24 hours (or as recommended by the manufacturer) for complete curing before testing the trap.

Testing of the trap can be performed easily with the use of sine waves. Begin with a continuous wave at the designed center frequency of the panel. You should be able to feel the vibration of the panel face. Slowly raise and lower the frequency to find the panel's actual center frequency. The center frequency of the trap will be the one where its vibration is greatest. You can use this information to help adjust panel design, if necessary.

Helmholtz Traps

Helmholtz traps are a type of membrane trap that does not rely on sealed spring cavities. There are many different types of these traps, some utilizing wooden slats, others using hardboard panels with holes drilled into them. We're going to focus on Helmholtz slot resonators in this book.

A slot resonator is constructed by using members to frame off the wall surface, similar to the panel traps, which then have insulation placed inside. The frame is subsequently covered with flame-retardant fabric, and finally slats of particular widths are placed with a slot (again of a particular width) between each piece.

The width of the slot opening is a critical piece of the puzzle when it comes to these working properly—with smaller slots, greater board widths, deeper slots, and deeper boxes lowering the frequencies. Thus, a disadvantage to this method is the painstaking attention to detail required to really get it right. A variance of even 1/16″ can drastically change the center frequency on these traps. However, these traps can also be constructed with a box depth that varies, making them a very good means of achieving broadband low-frequency attenuation.

Figure 9.12 gives you the formula for calculating a slot resonator. Once again, there is no deep math involved here. The formula is another easy one to program into a spreadsheet or to perform on a calculator (assuming it has a square root function key). There are also a ton of online calculators for these traps as well. Just do a search for "slot resonator calculators," and you'll find plenty of them.

Figure 9.12 Helmholtz slot resonator calculation.

$$f = 2160 * \sqrt{(r/((d*D)*((r+w)))}$$

Where:
f = Resonant Frequency in Hertz (Hz)
r = Slot Width
w = Slat Width
d = Effective Depth of Slot (1.2 times the actual slat thickness)
D = Depth of Box to the inside face of slat
2160 = c/(2 * π) rounded
c = Speed of Sound in inches/second
π = 3.14159

Be careful if you choose to go this route. Sometime, back a ways, there was an error in this calculation that made its way through the Internet. See Figure 9.13 for a side-by-side comparison of the correct and incorrect formulas to make certain the one you use is correct.

Figure 9.13 Helmholtz slot resonator formula correction[5].

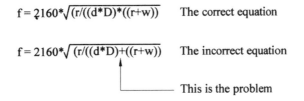

OK, on to how one goes about building one of these.

Figure 9.14 is a plan for a slot resonator. Note that one of the differences between this and a panel trap is that the insulation is placed touching the back of the slats. This helps to keep the slats from creating a vibrating noise when the panel is excited. Note also that you maintain airspace behind the insulation. This, too, is important for the trap to work effectively.

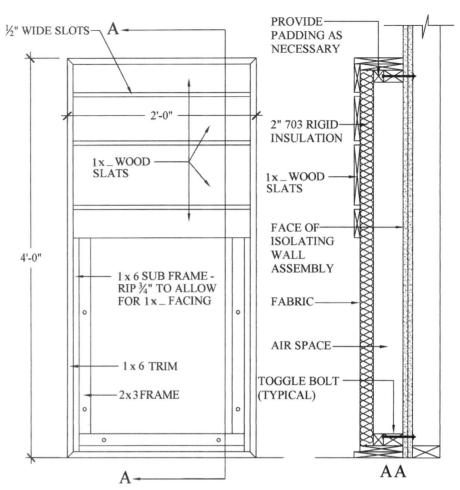

Figure 9.14
Helmholtz slot resonator plan.

NOTE: THE USE OF ALTERNATING 1 x 4 / 1 x 6 / 1 x 8 IS EQUAL TO THE USE OF 1 x 6 FOR THE PURPOSE OF CALCULATING THE CENTER FREQUENCY.

Figure 9.15 is an example of a broadband slot resonator built into a corner in a room.

This resonator is fairly broadband in nature and will handle frequencies ranging from about 66.44Hz to 155.82Hz. The estimated frequency range is based on the depths of the resonator—deepest point to shallowest point. Another method you can use to broaden the frequency range would be to vary the slot widths.

Once again, pay attention to the cloth being used for treatment. Guilford of Maine also has fabrics suitable for this work. Their style 2100 (for one example) is tested and approved for both FM 701 and ASTM E-84.

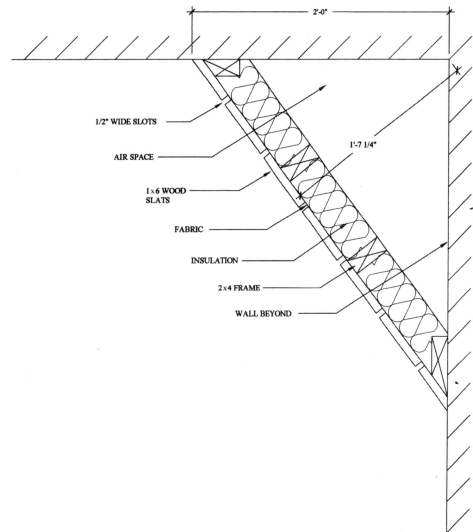

Figure 9.15
Helmholtz corner slot resonator.

Figure 9.16 is a 3D view of a slot resonator in the room corner. If the unit is constructed as shown (from floor to ceiling), it will be a reasonably effective unit. This is because it covers not only the corner (remember looking at bass buildup in corners earlier in the book), but also because it covers two trihedral corners. Recall that a trihedral corner is where a floor/wall/wall or ceiling/wall/wall meet, and it is the location of the greatest bass buildup in the space.

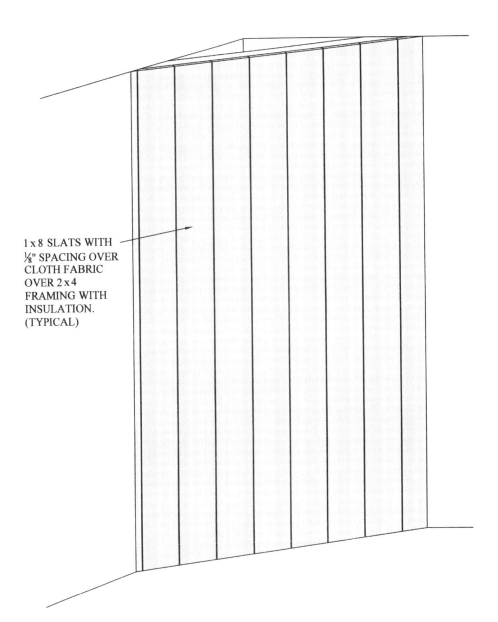

Figure 9.16
Helmholtz corner slot resonator isometric view.

1 x 8 SLATS WITH 1/8" SPACING OVER CLOTH FABRIC OVER 2 x 4 FRAMING WITH INSULATION. (TYPICAL)

Mid/High-Bass Absorbers

Mid/high-bass absorbers can be made to complement a series of low-frequency bass traps. Figure 9.17 is a plan for a typical 2'x4' mid/high absorber. These traps are manufactured similarly to the bass panel traps and will fit side by side with them, but they do not utilize a rigid panel for the face. Rather, this is a cloth finish and thus breathes air freely.

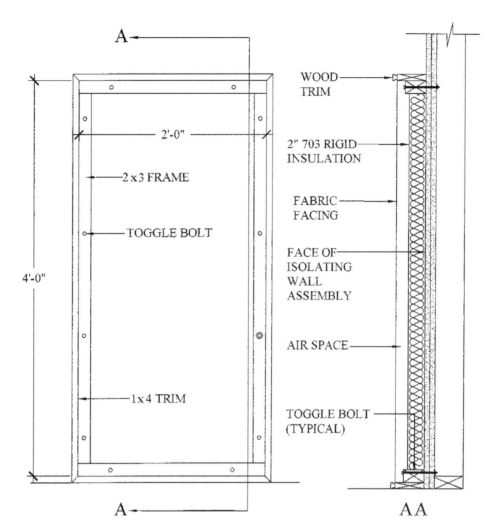

Figure 9.17
Mid/high-bass absorber plan.

Diffusors

Acoustic diffusion is created through a series of non-flat shapes that have enough dimensional variation to scatter the sounds that strike them. By the use of the term "non-flat," you should understand that this refers to flats in multiple planes, as well as non-flat shapes in general (for example, convex surfaces). Acoustic diffusion comes in quite a few different forms—the common pyramids and partial barrel forms, along with quadratic shapes.

Quadratic Residue Diffusors

In the 1970s, Schroeder's designs[6,7] were the stepping-off point for the widespread use of diffusion for acoustic control of room boundary issues. Schroeder presented methods of designing concert hall ceilings that could avoid direct reflections into the audience. In 1975, he provided a way of designing highly diffusing surfaces based on binary maximum-length sequences, and he showed that these periodic sequences have the property that their harmonic amplitudes are all equal.

Schroeder later extended his method and proposed surface structures that give excellent sound diffusion over larger bandwidths. This method is based on quadratic residue sequences of elementary number theory, investigated by A. M. Legendre and C. F. Gauss.

Dr. Peter D'Antonio[8] (aka Dr. Diffusor) took this to another level in the 1980s with his studies, and when he formed RPG Diffusor Systems and developed the Reflection Phase Grating Diffusor.

Suffice it to say that there is no easy way to break this down from a mathematical point of view. It took D'Antonio and Konnert fourteen pages just to describe the workings of the RPG Diffusor and that's done with the understanding that engineers were reading the paper—thus, they had no reason to simplify.

If this is something that you really want to pursue, I would suggest that you either purchase a kit (if you really want to construct it yourself), or turn to a company like Auralex or RPG to purchase something you know is going to work.

Figure 9.18 is a kit you can purchase from a company called Decware for their Model P1324 Quadratic Residue Diffuser.[9]

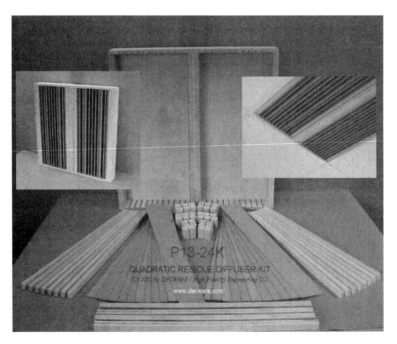

Figure 9.18
P1324 Quadratic Residue Diffusor kit.

Their product is designed based on a prime 13 sequence, which is repeated twice in their 24x24x3 inch diffusor. This gives you diffusion over three octaves—1,125Hz ~ 12,000Hz. These units are made from solid wood and are available in kit form (as shown in Figure 9.18) or assembled (as shown in Figure 9.19

Figure 9.19
P1324 Quadratic Residue Diffusor preassembled.

These modular units are designed to be used in groups. They can be fastened to wall surfaces, ceilings, or used with legs as freestanding units. They're a perfect fit in drop ceilings, as shown in Figure 9.20.

Figure 9.20
P1324 Quadratic Residue Diffusor in ceiling grid.

You can finish these by painting them, applying stain, or with clear finishes. In some cases, you can cover them with speaker cloth. To treat an area, no less than two units are recommended.

Polycylindrical Diffusors

A polycylindrical diffusor is simply a bent panel intended to scatter sound. It does this by taking the incident sound waves that strike it and reflecting them in many different directions.

In Figure 9.21, you can see a very simple representation of the scattering of sound striking a polycylindrical surface.

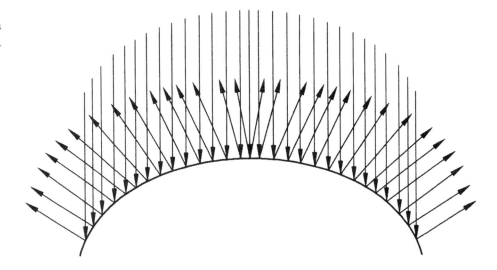

Figure 9.21
Sound diffusion from a polycylindrical surface.

For a DIY polycylindrical diffusor (also known as a "poly"), you take a sheet of plywood and bend it to create a surface that will diffuse the sound waves that strike it.

Making a poly is not nearly as difficult as you may think—with the biggest challenge being to make certain that the plywood you choose can easily bend into the shape you want.

When you go to the lumberyard, test the flexibility of the plywood you want to use and make certain it is capable of producing the arc that you want to construct. There is plywood that can be special ordered that is made specifically for bending—4'x8' sheets of $3/8$" lauan "barrel bend" run around $45 (USD) per sheet, while $1/8$" birch runs around $38 (USD) per sheet. This material can be bent to pretty much any shape you want. If you plan on building your own furniture, this is very handy for bent surfaces on that as well.

The easiest type of poly to construct is created using a simple elliptical arc to control the layout. Notice that I said "elliptical." Arcs that have a constant radius (and thus are circular in nature) are not the preferred method for poly design. An ellipse is a series of small arcs making up a larger arc. Although the ellipse may be symmetric about its horizontal and vertical axes, the left and right sides of the arc are made up of a series of varying radii. This creates a much better surface for diffusing sound waves than a simple arc can.

Figure 9.22 is a simple elliptical arc that you could use to serve as a poly.

Figure 9.22
Elliptical arc.

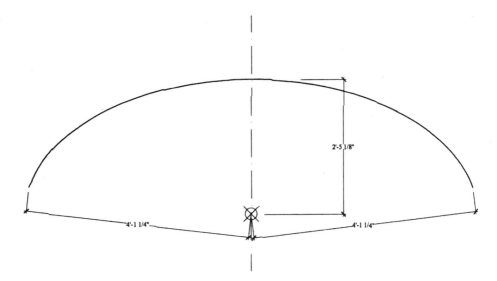

Developing the layout for a poly is not nearly as difficult as it might seem on the face of it. If you have a simple architectural drawing program, it's actually quite easy. Draw the shape you want to use in real life (meaning you do not "draw to scale," rather the drawing should reflect real-world dimensions) and then create a grid to control your layout.

In Figure 9.23, you can see an example of this with the same arc using a 3" square grid (which should be more than adequate to get the job done from the perspective of layout).

Please understand that the figures above are conceptual in nature and are not intended for you to use for the purpose of an actual poly design. The size and placement of polys are room dependent, although the basic rule of thumb would be to place them no closer to you than three times the wavelength of the lowest frequency you were looking to treat in this method.

If you don't have architectural software, it doesn't mean that you can't make a poly. You can still go from "A" to "B," although the lack of a working drawing will probably place some limits on you. However, you can always build a 1x4 box (use the same construction you see in Figure 9.2) and make it 47" wide by 96 $1/4$" long (those are clear inside dimensions), after which you take a piece of $1/4$" plywood, bend it along its length, and slip it into place. It will lock into place inside the frame, which will force it to bow out.

I would recommend that you have help when doing this, wear heavy work gloves, and use a couple of temporary cleats (placed inside the box frame) so that you create a free space for your fingers (just in case). You need to be very careful when placing bent plywood because a sheet of plywood (even $1/4$") could easily cost you a finger if you get one trapped between the frame and the edge of the plywood.

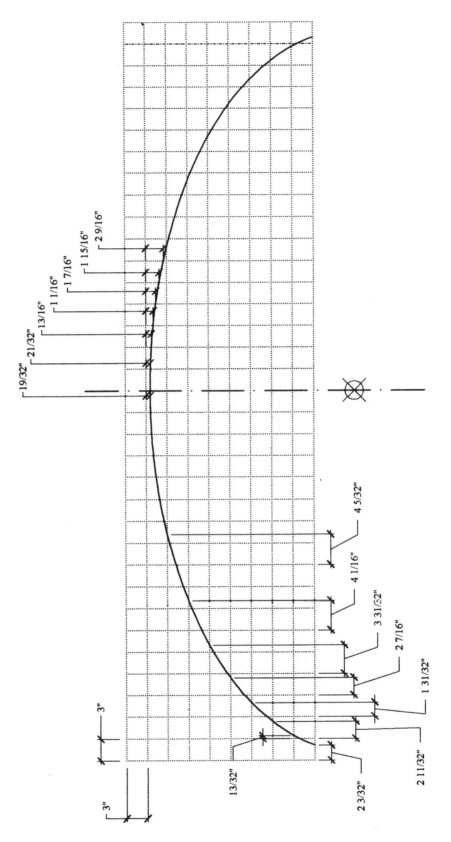

Figure 9.23 Arc layout.

You also need to understand that using arrays of equally sized "polys" can cause you more problems than you would have if you had not used them in the first place. Cox & D'Antonio have indicated that large arrays of "polys" are sources of comb-filtering effect.

The best approach is to vary the sizes (and orientations) of these treatments if large arrays of "polys" are your preferred approach for diffusion.

For a much more detailed reading on this subject, you could look through these papers/publications:

- ▶ ISO 17497-1:2004. "Acoustics—Sound-scattering properties of surfaces—Part 1: Measurement of the random-incidence scattering coefficient in a reverberation room."

- ▶ J. Cox and P. D'Antonio, *Acoustic Absorbers and Diffusers*, Spon Press, London and New York 2004.

- ▶ Bruel and Kjaer, DIRAC Room Acoustics Software type 7841, user manual. V. 4.0, 2007.

- ▶ "Measurement of the surface-scattering coefficient: comparison of the Mommertz/ Vorlander approach with the new Wave Field Synthesis method"[10].

Like everything else to do with acoustics, you could fill numerous books on this subject alone, and in fact many have been written.

Make the Best Use of Your Space

I see a lot of people designing home studios that have splayed walls with a ton of wasted space in areas that could (effectively) be used for low-frequency attenuation. Figure 9.24 is a plan view showing you one method you can use to take advantage of those wasted spaces.

Note that the isolating wall assembly continues through the space that would be wasted otherwise—almost like building a closet. The insides of this space are then fitted with corner absorption (on all four corners), as well as multiple layers of 4"-thick absorption installed the full width of the space, which has fluffy fiberglass insulation in between those layers to add even more to the package.

In Figure 9.25, you can see a vertical section through this assembly. Note here that there is no header assembly. The treatment runs from floor to ceiling to maximize the space available.

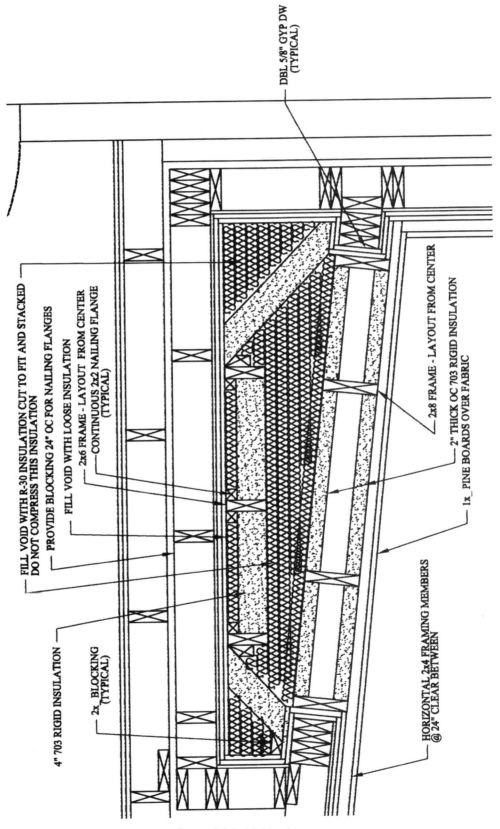

Figure 9.24 Hidden bass traps.

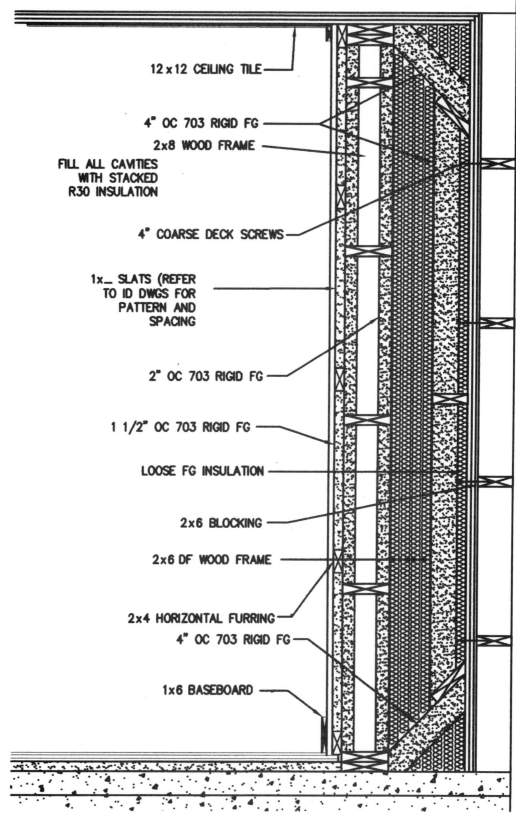

Figure 9.25 Hidden bass traps.

This is a very effective method for adding a whole lot of treatment into one small area.

You can also see that (in this case) the wall finish inside of the studio is exactly the same as it is on the remaining walls, making this treatment impossible to see. The wall treatment in this case is wood, but you could also use a cloth fabric (fire treated, of course) for the finish as well.

The back of the control room in Studio "H" in the Philippines was treated in a manner similar to this. In that case, a fabric finish with wood trim was used for the final room finish. You can view that if you look back to Figure 1.9.

Ceiling Clouds

This will be the last item I touch upon in the area of DIY room treatments.

In many cases, people have enough headroom that they are well able to afford losing a bit of it to a ceiling cloud.

Ceiling clouds are an excellent way to add a lot of room treatment without losing a single inch of floor space. The increase in absorbent material is a great way to help control low-frequency modal issues in a space, while (at the same time) leaving wall space open to create a more reverberant sound field.

Notice that I said "help control." This is because you typically will never get there (especially in small rooms) through the use of this treatment alone. It will also help you create a reflection-free zone between the floor and ceiling, and it will provide you with a place to install recessed lighting without the need to ever penetrate your ceiling membrane.

I have seen a lot of people install small ceiling clouds in various places throughout their rooms, but I prefer one large cloud, usually covering the entire floor with the exception of a (roughly) 2' strip around the perimeter.

Figure 9.26 is a reflected ceiling plan that includes an assembly creating an "acoustic cloud" that will be installed beneath the finished ceiling. In this design, the entire ceiling above the cloud is finished with a 12x12 acoustic ceiling tile.

The cloud will be hung from the ceiling through the use of eye-hook anchors penetrating through the finished ceiling, drywall, and into the structural members that create the ceiling frame. The frame of the cloud should have holes drilled through some of the inner frame members (about 1" above the cove molding), which will then have a length of 7x7 strand $1/16$" stainless steel wire rope looped through the hole with a twisted tie just above the top of the wood frame. The wire is then passed through the eye of the hook (that is attached to the ceiling) and again looped around itself to create a connection capable of carrying the load. Additional wires must also be installed to brace the cloud against movement under seismic conditions. This is a free hanging unit. You do not want it hanging above your head, able to move laterally.

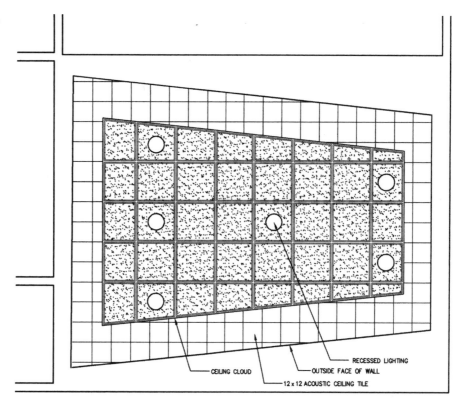

Figure 9.26
Reflected ceiling plan.

For the purpose of hanging a cloud like this, I seriously recommend that you hire a professional installer to do the job. They will understand the weights involved and the proper hanger to use, along with the proper methods of hanging and bracing the cloud.

Understanding Stranded Wire Rope

Good-quality wire rope is available with very small diameters that are capable of carrying rather large loads. A typical 7x7 stranded SS wire rope is capable of a safe load of about 480lb.

Wire ropes are indentified through the use of a numbering system. In general, the first number represents the number of wires that are wrapped to create a "strand," and the second number represents the number of strands that make up that particular rope. The greater the number of wires per strand and strands per rope, the more flexible the cable (and the higher the cost). Also, the larger the rope diameter, the thicker the wire is that creates the strands, and the greater the breaking strength will be.

Thus, a 7x7 wire rope would contain 7 wires per strand with 7 strands. A 7x19 rope would contain 7 wires per strand with 19 strands. The first would be less flexible than the second. They would both have the same minimum breaking strength with the same diameter rope.

One other comment: These products are usually made using either stainless steel or galvanized steel. Stainless steel will not rust, but it is not quite as strong as galvanized steel. Galvanized steel is a little stronger, but does not hold up to corrosion the way stainless steel does. With the size we are talking about here, you will not really see a difference in strength. However, I always get a little cautious whenever I am hanging something heavy, and I know that I will not see the hangers when it is all finished. So I would recommend you invest a little more for the stainless steel. (It is well worth it for the added peace of mind.)

Figure 9.27 is a couple of horizontal sections through that same ceiling cloud. It's detailing a typical end detail along with a section through a recessed light fixture.

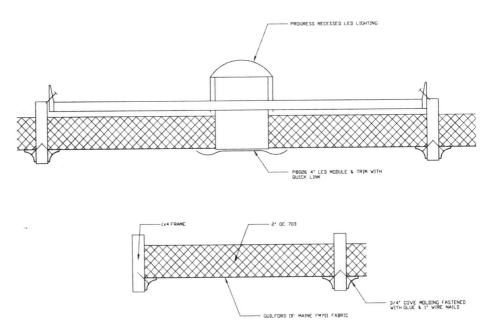

Figure 9.27
Section through ceiling cloud.

Manufactured Treatments

When discussing manufactured treatments, the two most popular methods of treatment are either high-quality acoustic foams or products utilizing specialized acoustic fiberglass. We are going to examine products here from a few different manufacturers, without any judgments being made (by the author) as to which one makes the best (or even better) products.

Because I know the products from these manufacturers and enjoy a personal relationship with parties in each firm that are involved in the design of these products, I am not showing you anything that I have not personally experienced. I do not, however, own stock in any of these companies, nor do I have any other vested interest in any of these firms.

These companies spend money on acoustic research and product development, and their products are at the forefront of product development in the industry. They are constantly looking for new ideas and methods for improving their products and the industry as a whole.

The products from these firms work. I've heard their effect in studios. I've examined before and after conditions using products from these manufacturers, and they all impress me. However, there are also other manufacturers out there who make similarly good products, and also some whose products I would not recommend. (I will not list those firms here.)

I do not list these other firms (who make similar products) not because their products are inferior, but rather because I do not have personal knowledge of their products. Thus, I can neither recommend nor oppose their products. As to the firms that I would not recommend, if you spend even a few minutes looking them over carefully, you will come to the same conclusions I did, which is why I do not need to name them. They have not invested the time or expense to have their products tested at a recognized facility, and therefore cannot back up any claims to their products' performance.

When looking at companies who have claims that test data proves their products to equal those shown here, look carefully for the actual test reports and request them if they are not published. If they cannot (or will not) produce reports for you, or if those reports are not from a recognized, reputable testing laboratory, then do not even consider their products. Any reputable manufacturer will provide copies of the actual test data to make the sale (if that's what it takes).

Please understand that testing in a manufacturer's "special testing facility" is not the same as testing at an independent certified laboratory for a variety of reasons, the least of which is that (without suggesting that anyone might actually tell an "untruth") certified labs have a vested interest in publishing exactly what information they observed during the testing procedure. This would include "the good, the bad, and the ugly." It would also include exactly what the test procedure was, and if it varied in any manner from the standards they were testing to.

You can take this as an absolute: No testing lab would ever even consider falsifying the data they report, regardless of some people's claims that they will protect the best interests of the clients who are paying for the tests. They will always protect their own best interest, which is keeping their certification so they can stay in business. If they were ever caught falsifying data, they would stand to lose that certification, which would mean they would not have the right to bid on very lucrative government contracts, which make them a whole lot more money than any single manufacturer is ever going to pay them. They would also lose any credibility in the marketplace.

Another point on that subject is that you have no way of knowing whether the construction of that manufacturer's facility comes even close to the standards required by ASTM or ISO, which means that you have no way of knowing if the testing in their room might skew the test results in their favor.

Also, beware of companies who publish reported testing results but fail to produce copies of the actual reports upon request. I had one company fail to reply to my request for test data, and only upon a subsequent request from me for the data (along with a comment that I found it very interesting that their absorption coefficients were exactly the same as those in a report I had received from one of their competitors) did they finally respond. I found the response both interesting and enlightening, it was that they had never actually had their products tested. Apparently, they believed that their products were comparative to the company whose report I had, and thus decided to save the money on testing and just used those numbers for their purposes. After some more communication back and forth, they finally agreed to have their products tested. Imagine my surprise to find that their products did not compare with their competitor's. I cannot even begin to picture how many unsuspecting customers they sold these products to.

If you decide to shop for yourself with unknown firms, then remember the old saying: "Caveat emptor" (*Let the Buyer Beware*). Because in the end, the only person really responsible to protect you from getting ripped off is you.

The first edition of this book contained information on all of the products carried by all but one of these manufacturers, including copies of the data from the test reports I received from each of the companies. This edition, however, will only feature a few of the products and none of the test data. You can obtain the data directly from the manufacturer, and by now you should understand enough about acoustic coefficients (and sabins) to make an intelligent, informed decision regarding their suitability for your purposes. I say "all but one" because I have added a company that was not included in the first edition.

Auralex

The two most commonly used sound-absorption materials (for manufactured treatments) are high-quality acoustic foam and specialized acoustic fiberglass. Generally, acoustic foam is just referred to as *foam*, although there are some very dramatic differences in cell structure and density between acoustic foam and the thousands of other types that are manufactured. This is why you can't just buy mattress pads with which to treat your studio acoustically.

In addition to the two most popular types of acoustic absorption materials, Auralex offers a Class A, fire-resistant, natural fiber panel called *SonoFiber*. SonoFiber acoustic panels are the perfect solution for those budget-conscious projects requiring a Class A fire rating without the aesthetic demands of designer treatments such as fabric-covered panels.

Wood is also a beautiful option for room treatments (diffusion primarily).

Auralex manufactures products made using all of these options. Here are some really nice choices.

SpaceArray

For audio professionals seeking to maximize the acoustical performance of their recording spaces, pArtScienceSpaceArray diffusors disperse sound waves evenly and randomly to provide a consistent acoustical environment in any room. The SpaceArray diffusor employs a quasi-random array using state-of-the-art engineering techniques and carefully selected, high-quality materials for superior sound diffusion. It has the following qualities:

- ▶ **Outstanding performance:** Eliminates flutter echoes and other acoustical anomalies without removing acoustical energy from the space.

- ▶ **Quasi-random array:** Randomization of reflections evenly distributes sound in a space, taking the guesswork out of microphone placement and mixing. It just sounds better.

- ▶ **Modular:** 2' by 2' panels used in a variety of applications and placement options, including "T-Bar" grids.

- ▶ **Beautiful solid-wood construction:** Widely used for musical instruments and decorative finishes for over 1,000 years, Paulownia wood has one of the highest strength-to-weight ratios of any wood in the world.

Figure 9.28 is an example of this diffusor.

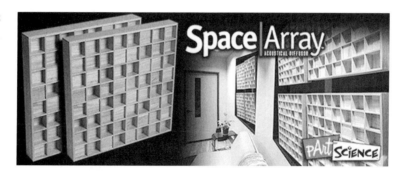

Figure 9.28 SpaceArray diffusors.

Space-Coupler

The pArtScience SpaceCoupler is an acoustical treatment that creates a natural "large sound" within a small room footprint. Unlike current alternatives, which involve custom design and remodeling, the SpaceCoupler works within the current room footprint for a fraction of the cost. Here are some of its features:

- ▶ **Loosely couples spaces:** Creates a natural "large sound" in an otherwise small room.

- ▶ **Waveguide design:** Redirection of sound energy offers an attractive alternative to traditional absorptive and diffusive surface treatments.

- ▶ **Modular:** 2x2 panels used in a variety of applications and placement options, including standard "T-Bar" ceiling grids.

These units are made with the same beautiful wood used for the SpaceArray diffusors. Check out Figure 9.29 to see this product.

Figure 9.29 SpaceCoupler.

Both the SpaceCoupler and SpaceArray were designed by Auralex in conjunction with one of the leading studio designers, Russ Berger. Russ is the president and acoustical design principal of RBDG (Russ Berger Design Group) and also currently serves as president of the National Council of Acoustical Consultants (NCAC).

AudioTile

For audio professionals looking for aesthetic alternatives to acoustical absorption treatments, pArtScience AudioTile ShockWave delivers maximum broadband absorption for a pleasing, well-controlled sound without being perceived as too dry. Unlike traditional products, the patent-pending AudioTile offers unlimited design possibilities for one-of-a-kind personalization and a custom designed look. Here are some of its features:

- ▶ **Maximum broadband absorption:** 0.81 absorption coefficient at 125Hz. Max. NRC = 1.00 to 1.05, depending on installation.

- ▶ **Varying thicknesses:** Foam thicknesses from 1" to 4" are built into each pattern, providing varying degrees of absorption at varying frequencies. Greater thicknesses can be achieved with alternative layouts.

- ▶ **Unlimited tessellation design patterns:** Allows a unique means of blending absorption, diffusion, and reflection.

- ▶ **Easy Installation:** Mount by using Foamtak spray adhesive, Tubetak Proliquid adhesive, or Temp•Tabs Studiofoam mounting kits.

Figure 9.30 is an example of the AudioTile ShockWave

Figure 9.30 AudioTile ShockWave.

Auralex also carries products for sound isolation and some attenuation products not listed here. Please visit their Web site.

Auralex Acoustics, Inc.
6853 Hillsdale Court
Indianapolis, IN 46250
PHONE: 317-842-2600
FAX: 317-842-2760
http://auralex.com

RealTraps

The RealTraps product line is based on membrane bass traps that also absorb mid and high frequencies. This makes them a total solution for recording studios, listening rooms, home theaters, auditoriums, and anywhere economical, yet where very high performance acoustic treatment is required.

All RealTraps panels may be hung from the walls, installed straddling corners, mounted on inexpensive microphone stands, or mounted on RealTraps stands.

All of their products are small and lightweight, easy to handle, and can be shipped economically. They mount easily with one screw or hook, just like a picture, without glue or permanent wall damage. They can be mounted vertically or horizontally as space permits, or installed at the top of a wall where they're out of the way. Since RealTraps products are made with rigid fiberglass and metal, they're non-flammable with a Class A fire rating, so can be installed with confidence in public venues.

MiniTraps

MiniTraps are 2'x 4', 3 1/4" thick and weigh 18 lb. Despite their small size, these are real performers that have exceptional specs, especially at low frequencies. Figure 9.31 is a shot of this product in the control room at Nile Rodgers' studio.

Figure 9.32 is a picture of this product installed in a living room that also serves as the owner's piano room.

Figure 9.31
MiniTraps in Nile Rodgers' control room[11].

Figure 9.32
MiniTraps in living room.

The Real Traps Diffusor

The Real Traps Diffusor is a very attractive example of a marriage between a bass trap and a QRD well-type diffusor. It transitions from diffusion to absorption over the range between 400 to 800Hz.

Figure 9.33 is a beautiful shot of this product.

Figure 9.33 Real Traps diffusor.

More Information

RealTraps co-owner Ethan Winer is well known in the industry for his many articles in popular audio magazines, and all of his recent articles about acoustics and room treatment are available on the RealTraps site.

RealTraps
34 Cedar Vale Drive
New Milford, CT 06776
860-210-1870
www.realtraps.com

GIK Acoustics

When the first edition of this book came out, GIK Acoustics was a small company just getting its feet off the ground. Today, they have expanded their operations and now have a sales office in England.

They have also expanded their product line, which now includes diffusion (both Polys and QRD), fancy artwork facings for their panel (using your artwork if you choose), and even attractive table-top furniture with treatments concealed within. All in all, they have become a "player" in the field of room attenuation products.

GIK 244—2x4 Acoustic 4" Panel

Check out Figure 9.34 for a sample of this product. The GIK 244 is manufactured using 4"-thick, 8pcf acoustic mineral wool. The panels are fitted inside a 5 1/2"-deep wood frame so they can be mounted tight to a wall while maintaining 1 1/2" of airspace behind the insulation.

Figure 9.34
GIK 244 acoustic panels.

GIK 242—2 x 4 Acoustic 2" Panel

The GIK 242 panel (Figure 9.35) is manufactured for the purpose of creating a reflection-free zone for the listening position in the room. The GIK 242 panels are manufactured using 2" 8psf mineral wool mounted inside of a 2"-deep wood frame. Additional benefits can be had by mounting these panels off the wall.

Figure 9.35
GIK acoustics 242 acoustic panels.

The GIK QRD Diffusor is a quadratic root diffusor based on a prime 7 sequence. This product is just 19.5"x 45" and 5.5" deep. It's an attractive package that will fit easily in most spaces. It is available in a variety of colors, and the depth is not going to cost you a whole lot of space. Figure 9.36 is this package with a black interior and a dark-stained finish on the wooden frame.

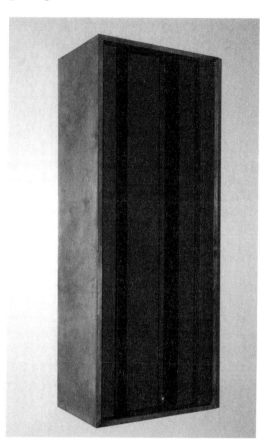

Figure 9.36
GIK QRD Diffusor.

More Information

If you think that this company may have what you need (or perhaps even just want), please check out their Web site.

GIK Acoustics
404 486 8996
Fax 770 234 5919
www.gikacoustics.com

Ready Acoustics

Ready Acoustics is a company that began by providing very high quality bags designed to fit acoustic material in (rigid fiberglass or rockwool). This was intended to provide a cost-effective option for DIY treatments for people who were unsure about their ability to produce a professional-looking fabric wrap on their own.

From there, they grew and now offer (in addition to their line of DIY products) a full line of room treatments. These are very attractive products and use some interesting designs along the way.

Figure 9.37 is a product from them called the Chameleon Bass Trap.

Figure 9.37
Chameleon Bass Trap.

Figure 9.38 is a picture of their products in a tracking room at the Monument Sound recording studio.

All in all, they have a really nice line of effective products, and they still carry a full line of DIY products, if that's the direction you should decide to take.

Figure 9.38
Monument Sound recording studio.

More Information

For more information regarding this company and their product line, please visit their Web site.

Ready Acoustics LLC
PO Box 626
Elk River , MN 55330
Phone: 800-820-5044
Fax: 866-320-1800
http://www.readyacoustics.com

So Where Do We Go from Here?

By now you should understand the basic concepts of acoustic isolation, studio electrical needs, the special needs of HVAC and fresh air (as it pertains to studios), door and window design, room treatments, and room testing. All that's really left is to figure out how to put this all together. That is where the next chapter is going to take you.

One comment before you flip the page and get started (reading Chapter 10 that is), Chapters 11, "Myths and Legends," and 12, "Codes, Permits, and Special Needs," although they do not specifically have to do with the design and construction of your studio, deal with some issues and questions that I felt were important enough to include in this book. Please make it a point to read them before you get started on your design, because they really do have information that you may well want to consider as a part of the quest you are on.

Endnotes

[1] Provided by Jeff D. Szymanski, PE printed with permission.

[2] Provided by Jeff D. Szymanski, PE, printed with permission.

[3] ASTM C 423, Standard Test Method for Sound Absorption and Sound Absorption Coefficients by the Reverberation Room Method

[4] University of North Carolina at Chapel Hill, "How Many? A Dictionary of Units of Measurement" Russ Rowlett PhD,

[5] This error was first noted, and reported by Scott Smith, Newsgroups: alt.sci.physics.acoustics, on 02-02-2004 and was further verified by Eric Desart on 02-19-2004 at http://forum.studiotips.com.

[6] Schroeder, M.R., "Diffuse Sound Reflection by Maximum Length Sequences," *J. Acoust. Soc. Am.,* Volume 57, No. 1, pp. 149–150, January 1975.

[7] Schroeder, M.R., "Binaural Dissimilarity and Optimum Ceilings for Concert Halls: More Lateral Sound Diffusion," *J. Acoust. Soc. Am.*, Vol. 65, pp. 958–963, 1979.

[8] D'Antonio, Peter; Konnert, John H. "The Reflection Phase Grating Diffusor: Design Theory and Application," *J. Audio Eng. Soc.*, Vol. 32, No. 4, pp. 228–238, April 1984.

[9] Photography by Decware, printed with permission from Decware.

[10] A. Farina, Industrial Engineering Dept., Università di Parma, Via delle Scienze 181/A 43100 PARMA, Italy

[11] Nile Rodgers, Le Crib Studios in Westport, CT

[12] Monument Sound Recording Studio, Monument, CO.

CHAPTER 10

Putting It All Together

In this chapter, you're going to work with studio design—anticipating treatments and looking for design problems that are going to cost you something big down the road. You'll learn how to think your way through a design from start to finish, using information you've gathered from the various chapters of this book along the way. Don't hesitate to go back to those chapters for reference as you work your way through this material.

This chapter is also the one you want if you're living in a condo or apartment and can't alter the structure, but need to do treatments. If you fit into that category, then skip down to the subheading, "Studio Treatments," and begin your reading from there. (Unless, of course, you just find this so fascinating that you have to read it all.)

If you are going to design a studio for yourself, you'll examine situations that work and situations that would have worked except…oh well.

Yes, you'll examine some options in here that could have better outcomes if you're careful. As has been pointed out to you throughout this book, it's pretty sad when you get to the end of a project and have to say, "I forgot _____." (Let's hope you never have to fill in that blank.)

This chapter is going to walk you one step at a time through the entire design process and show you how to think your way through what's facing you. It will help you develop a feel for how you need to plan and what's involved with each step of the process.

Please take the time to study this chapter carefully, because it will make the difference between having "smooth sailing" through the construction process or spending a long, long time trying to work your way out of problems. A few extra days, or even weeks, of planning on your part is going to save you more than that in productivity (and money) in the long run. The intent here is not to design a studio for you (although if something here works for you, then by all means feel free to use it), but rather to learn to think your way through the process. It's also meant to teach you how to look at something you want to build, i.e., picture the pieces involved and then determine how you will assemble your studio at each particular location when you come to it. The end goal is a studio that you can be proud of and that gives you everything you paid for, namely, isolation with good acoustics.

Obviously, this book cannot cover every possible condition that you might encounter, nor will it try to do so. In fact, it shouldn't have to. Each of the conditions you'll examine will also work with other situations; it's all a matter of scale. For example, if you have a sanitary pipe that runs below your ceiling against a wall, the detail for a duct chase below the ceiling will work for your needs. Just scale it down to meet the size of the pipe. By the same token, should that pipe run through the middle of the room below the ceiling, the detail for the dropped beam will be perfect to enclose it. Again, it's just a matter of scale.

Remember, this exercise is intended to help you think your way through the process. So lay out your room, and then take a deep breath and look at each item in the space that will affect you. Determine beforehand how you will deal with that particular item; then put it down on paper and study it. Make sure that you have taken all of the pieces that make up that item into consideration. Like the pipe in the room—does your plan take into account the hangers or the bell size? It won't do you any good to design and construct your chase to fit the pipe and then have to thin out your drywall to accommodate the fittings.

Yes, it's going to take you time to work your way through this, but hey, you've waited this long to have your studio, so an extra week or two of proper planning isn't going to kill you.

Let's begin by taking a look at a layout inside of the basement of a 24x42 ranch.

Studio Design and Detailing

Begin by planning to keep your surround walls a minimum of 1" clear of the inside face of the foundation. This should allow the walls to be completely free from touching the existing structure, even if there are slight variations. If you have an older stone foundation—one that wasn't built with anyone worrying about the inside face being nice and true—the best you'll be able to do is to stay as close as you can.

Figure 10.1 shows you a typical layout of a ranch basement, with a center stair that's built off the face of the carrying beam, which is typically constructed at the centerline of the house lengthwise.

Hopefully, your electrical service, boiler (furnace?), and washer/dryer are all located on one end of the house, which gives you the opposite half of the basement for your studio. If not, you wouldn't want to move your boiler, but it shouldn't be a killer of a deal to relocate your washer/dryer. And, if your electrical service is located in the other half (your half) of the basement, you can live with this.

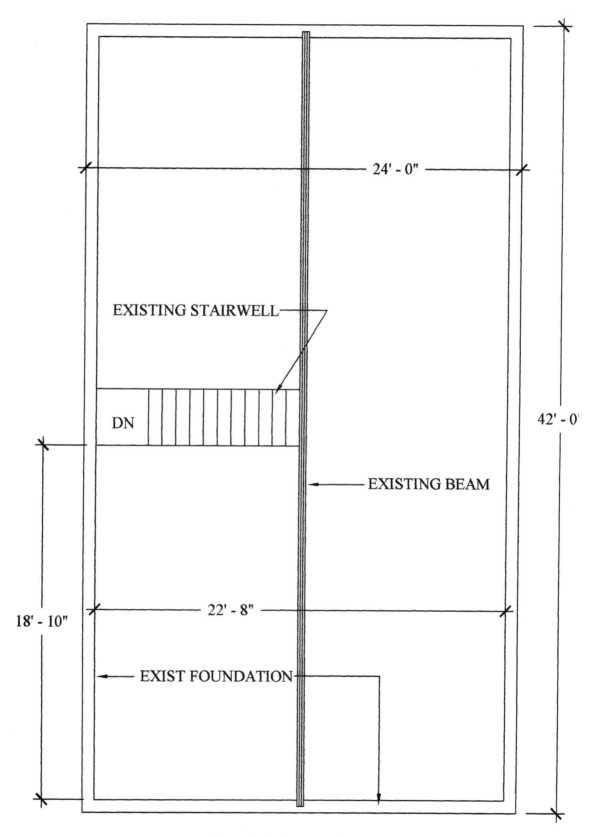

Figure 10.1 Typical ranch basement.

One of the things that will be important during the design phase of the work is that you delineate your existing structure accurately by creating as-built drawings. That means that you really do want to know exactly where your floor joists are in relation to the real world. The same goes for structural beams, columns, piping—anything that might interfere with your plans during the construction process, etc.

Figure 10.2 shows an example of an as-built of the deck and structural support members for this home. This is an exercise well worth doing. You'll use this information later in the process.

There are a lot of fairly inexpensive and free CADD programs out there. CADD programs are architectural drawing programs that let you accurately put in the computer what exists in the real world. The learning curve for these isn't all that bad. It would be worth your while to find one and play with it until you're reasonably proficient. You'll be glad you did in the end. You do have the option of doing this the "old-fashioned" way—drawing with pencil and paper—*drafting* is what we used to call it.

For the sake of argument, let's assume that the bottom half of this drawing is your half, which would give you 22' 8"x18 10" of space to work with. What exactly are your options? Well, that's 427s.f. of space, which would allow you enough room to build either a great combo room or a comfortable little multi-room studio containing the following:

- Iso-booth—32s.f.
- Control room—101s.f.
- Main room—180s.f.
- Cord/mic/gear storage—21s.f.
- Vestibule to control room—19 $^1/_2$s.f.

OK, now you're thinking that something must be wrong. Those numbers only add up to 353.5s.f., so what happened to the other 73s.f.? Well, as with everything else in the world, "We deals with what we gots ta deal with." The advantage to multiple rooms is isolation while recording. The disadvantage is that all those isolating walls eat up a lot of good square footage of real estate. In this case, that would be the equivalent of a room that's slightly over 8x9 in size.

Picture that with double-wall assemblies and a 1" air space, you have two rows of 3 $^1/_2$" studs, plus the 1", plus at least four layers of $^5/_8$" drywall total, so 10 $^1/_2$" of total wall thickness. And if you add another layer of drywall on each side, then 11 $^3/_4$".

That's almost one square foot of floor space for each linear foot of wall.

Also, a single room wouldn't have the entire area available either. You still need the perimeter wall (the one adjacent to the foundation) and a double assembly by the stairway, which would give you 385.5s.f. of clear floor space with one open room, so the multiple rooms only really cost you 32s.f. Let's look at the pros and cons for each approach.

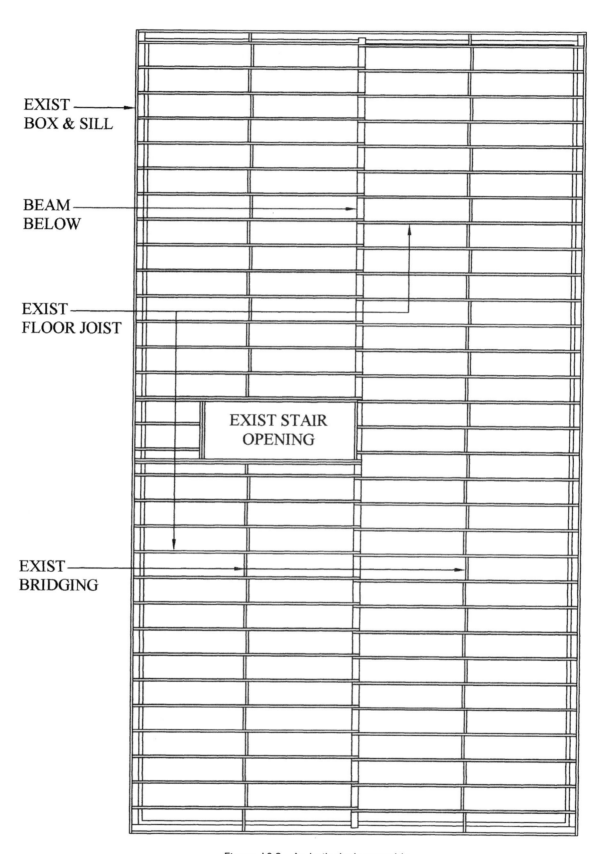

Figure 10.2 As-built deck assembly.

Pros for an open combination studio/control room:

- Larger room size generally means minimizing the amount of room modal activity. Remember that the smaller the room is, the easier it is to excite modes.
- A larger room allows more usable wall space and floor space to set up gear, which means much fewer windows and doors.
- If you're recording yourself (or as a part of a band), then you're in the same room as the recording gear, which makes control of the gear easier.
- If this room doubles as your practice room, a larger space is always more comfortable.

Cons for the same room:

- You won't have any isolation during the act of recording, and even with headphones on, bleed-through will color the mix.
- It's difficult to isolate vocals from other microphones in the space, and any vocals captured from instruments' microphones make it difficult to re-dub vocal sections.

Pros for separate rooms:

- You can isolate vocals from the band more easily.
- Monitoring a live recording can be done without the need for headphones.
- It just looks more professional, and if you become proficient at this, you might want that professional look if you pick up some work with local bands.

Cons for separate rooms:

- If you are part of the recording (i.e., guitarist, drummer, etc.), then running back and forth to turn gear on/off is a pain, and a remote control for the recorder becomes a necessity.
- It's difficult to set up and monitor levels, etc., if you're not near the gear (once again when recording yourself).
- A smaller room means greater modal activity. The smaller the room, the more treatment will be required to correct for this.

Now, this is all a personal thing. For example, I prefer a single larger multipurpose room. Not that I haven't thought about just how nice it would be to have a "real" studio, but my real estate dictates what I can and can't do.

In a multipurpose room, if you don't mind monitoring with headphones, you can set your vocalist's levels so that she/he can almost whisper into the mic, eliminating the problems with bleed-through into other mics. Just cut the vocals in later. Besides, if you're accustomed to being recorded, you're used to wearing headphones anyway.

You can DIY the bass and guitar for the initial recordings if you choose, so you can just capture the percussion (with mics on the room) and dub the other instruments back in later. But, in the process of this, you also tend to lose some of the spontaneity that takes place when recording a band live.

So with a series of smaller rooms (still large enough for "the band"), you can set up some gobos, record live, isolate your vocals, and just go for it. Ultimately, it's your decision to make.

On to Options

For the purpose of this exercise, let's take a look at a multi-room space constructed within your basement, based on the original basement you looked at in Figure 10.1. Figure 10.3 is a layout of a multi-room studio set within the space. Not a bad little layout—nice control room, small vocal booth, main room large enough for a four-piece band.

By the way, the drum set you see sitting by in the corner of the main room is used in this chapter just to give you a decent perspective of room size. It's a fairly large kit—24" bass—14" snare—toms are 8",10", 12", 14" (mounted) with a 16" floor tom. Brass includes a 14"HH, 14" crash, 10" splash, 12" splash, 18" crash, and a 20" ride.

This is drawn accurately to scale, so you can picture what else fits into the space.

So you just finished designing this, brought it to an acoustic Web site for review, and are awaiting comments.

The comments come back as follows:

> "The concept is nice. I like the fact that you've maintained symmetry in the control room. The storage room is a good use of a space that probably would be wasted otherwise and works well with creating that airlock entry into the control room for added isolation."

> "The same goes for the airlock into the room—a little extra isolation from the remainder of the house. Mom can do laundry while you're down there making music."

> "But…" (don't you just hate it when that happens—when someone says "but" after saying all those nice things?) "I don't think that you've thought through all of your other needs with this layout."

> "For example, your HVAC requirements."

> "Let's take a few minutes to work this through. You obviously need to learn to think like a designer."

Home Recording Studio: Build It Like the Pros, Second Edition

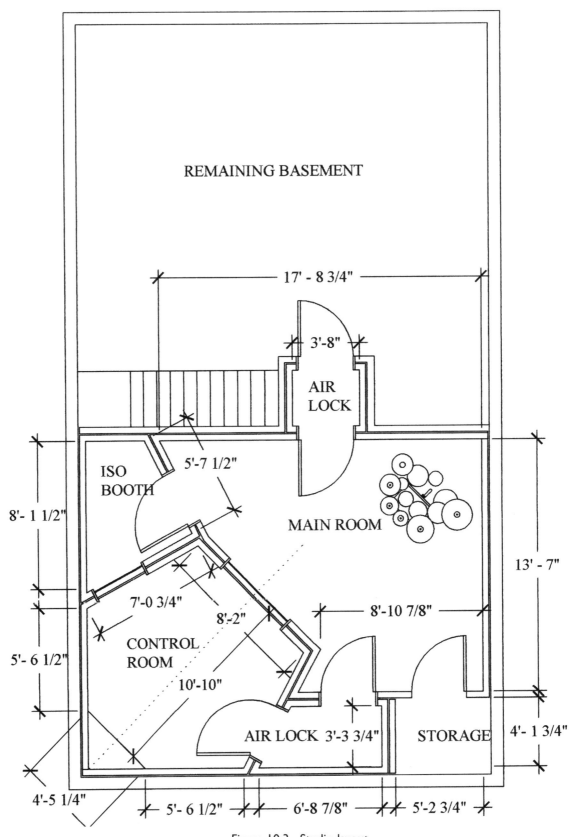

Figure 10.3 Studio layout.

Picture for a moment here what would be involved with getting HVAC ducts from the area above the stairway to the control room and vocal booth. If you tried to do a duct run through the space at the bottom of the stairs, you would find that it creates a headroom problem, and you would not have code-compliant stairs when you finished. This would be a bad thing.

Ducts tend to take up quite a bit of space, and for that reason, you want them to stay away from running through the bodies of your rooms to the best of your abilities. Take a look at Figure 10.4. This is what you would end up designing for duct runs with this layout.

Look carefully at the way the duct runs through the space. At the right-hand wall of the main room, you have a return air duct run, which requires a chase to box it in. Note that in order to get enough fresh air back into that duct from the main room, you'll have to create the funny little jog you see down by the storage room. Plus, the supply duct has to poke out into the room (right above the entrance door) far enough so that you can create the takeoffs from that to the rooms adjacent to the main room. So you'll have this funny little "box out" that you'll have to build below the ceiling, which exists for this duct as well.

You can provide a push (of air) across the room from the end of that duct as shown, but that leaves a couple of dead spots in the room that won't receive air flow, so you have to add some supply registers to the left and right of the main plenum as well.

The supplies for the iso-booth and main room can run above the ceiling until they hit the left-hand wall, and then the duct has to drop down below ceiling height to continue. This requires a duct chase in both the iso-booth and control room. You can (as shown) install top takeoffs from that supply duct, so that the remainder of your duct in those rooms is above the ceiling.

Your total duct material requirements within the room itself will be the following:

- Main plenums—supply and return—25' 10"
- Branch duct runs—91' 9"
- Supply registers required—8
- Return grilles required—5

That's quite a bit of materials for this small space. Let's see if we can shorten that list up a bit and clean up the rooms at the same time.

One other comment regarding this layout. The beam that carries the floor above cuts directly through the window and door openings into the control room. While the window is located in such a position that there can be a work-around developed for this (more on that a little later in the chapter), the door is another story.

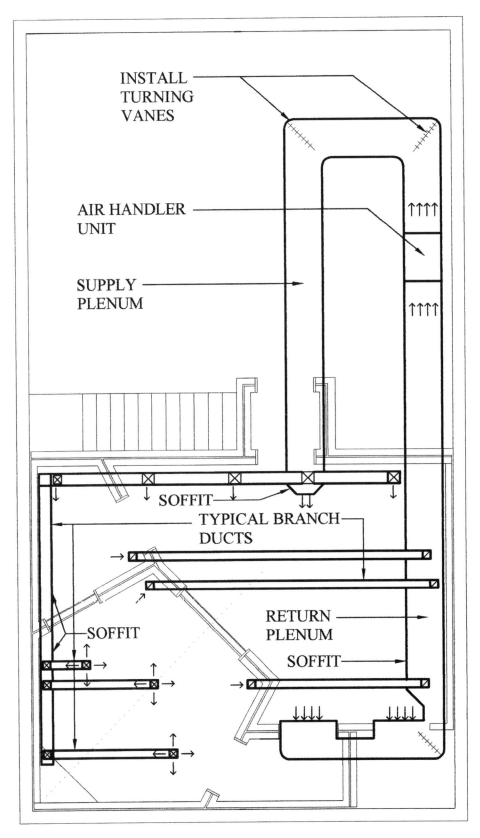

Figure 10.4 HVAC duct layout.

The beam cuts directly through the jamb on one side of the door and does not allow you enough room to install all of the isolation required. Picture that on the door side of the beam, you have to install framing and drywall, which can't touch the beam itself or you lose the isolation between the room and the existing building's structure. Also, for the door, you need gasketing and the door itself to be free and clear of the beam, not to mention that you also need a door jamb (frame). None of this is possible with this beam. Thus, you are going to need to find a way to install your door without the beam being a part of the equation.

Here are the options for this situation. Working within the same footprint, you can try flipping the room layout around to see if something might work. Since you worked very hard to get a layout you liked, why throw it away if you don't have to? Because it fits within the space, there are four possible layouts that work with this floor plan. Let's look at one of them.

Figure 10.5 is basically the same layout, but oriented differently within the space. This is simply a mirror image of the original plan drawn on the W-X axis of the room (left to right) directly through the room center, with the exception that the door into the control room has been rotated so that it's completely free from the beam. This also required a modification of the wall adjacent to the iso-booth to maintain symmetry in the control room.

What about the other options? Well, one of them you already worked through and discarded due to issues. Another would be to mirror that layout with a flip from left to right, but that would leave you in the same position as the original. The third would place the studio entrance through the control room, which would be less than desirable from both the entrance point of view and the lower ceilings (in the control room) due to duct runs. So the option we're examining now is really the best of the four.

Note how the gear closet and airlock now create a passageway into the space for the main HVAC supply plenum. If the ceiling heights in those spaces are lower than the main rooms, it is not a big deal. Let's add some ductwork to this to see how you make it from "A" to "B."

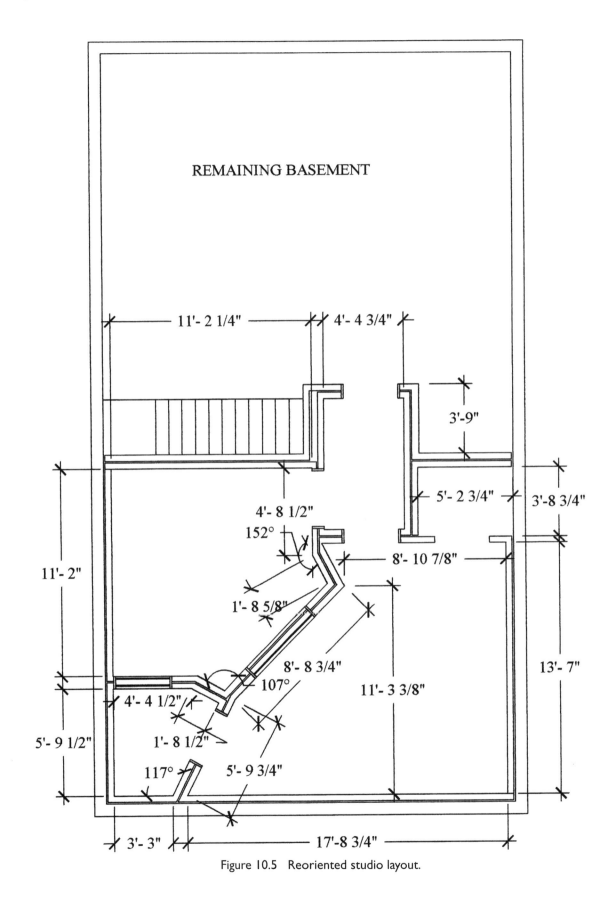

Figure 10.5 Reoriented studio layout.

HVAC Systems

Figure 10.6 adds a duct layout for this space. The air handler is still sitting in the house side of the basement. By the way, make sure to use isolation hangers for that equipment. The main room now has a duct chase (for the main return plenum) located at the right-hand side wall only. The main supply ends inside of the entrance air lock.

You will have a couple of small soffits required in the control room. None are required in the iso-booth, and the body of the main room is clean.

Note how the lateral runs with this layout all step up inside the space within the existing floor joist. This maximizes the ceiling heights in your studio. You can tell this because they show up as top takeoffs from the main plenums.

Note also the layout of the supply/returns. In the main room, you have one supply that ends at the face of the wall for the air lock entering the room and one small branch to the right. The two supplies that you needed on the left are now gone. On the opposite side of the room are a smaller series of return grilles (located in the ceiling), which are intended to draw the air evenly through the space. The supply in the iso-booth ends at the face of the wall, while the return grilles are located in the ceiling.

Your new total material requirements within the room itself are the following:

- Main plenums—supply and return—31' 0"
- Branch duct runs—78' 10"
- Supply registers required—6
- Return grilles required—6

With this layout, you require an additional 5' 2" of main plenum; however, you save 12' 11" of branch duct. Plus, you save two supply registers, which are more expensive than the one extra return grille that you need with the new plan.

The layout in the control room (in either case) is perfectly symmetrical within the space. It's just hard to envision with an angled drawing. Create a reflected ceiling plan (as if the floor were a big mirror), rotate the room 45°, and you'll see it clearly.

Figure 10.7 is a reflected ceiling plan (RCP) of the studio.

Remember that inside the control room, symmetry is everything. The HVAC layout here gives good air distribution throughout the room and places air directly above you and your gear. You'll appreciate this feature when it's hot outside. The return air draws to the front of the room, pulling the heat that you and your gear give off away from you.

You can see the soffits developed to enclose duct runs. Note in the control room that there are added soffits to maintain the symmetry of the room, the balance. When it comes to treating the room, these can be used for bass traps.

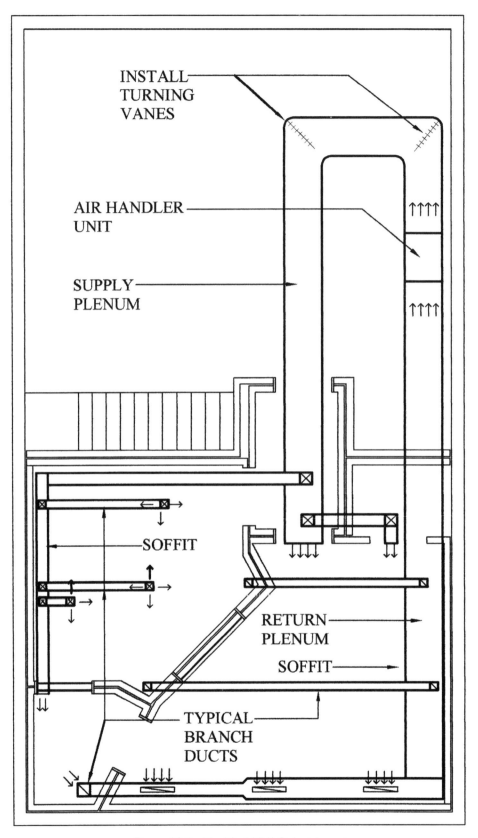

Figure 10.6 Modified HVAC duct layout.

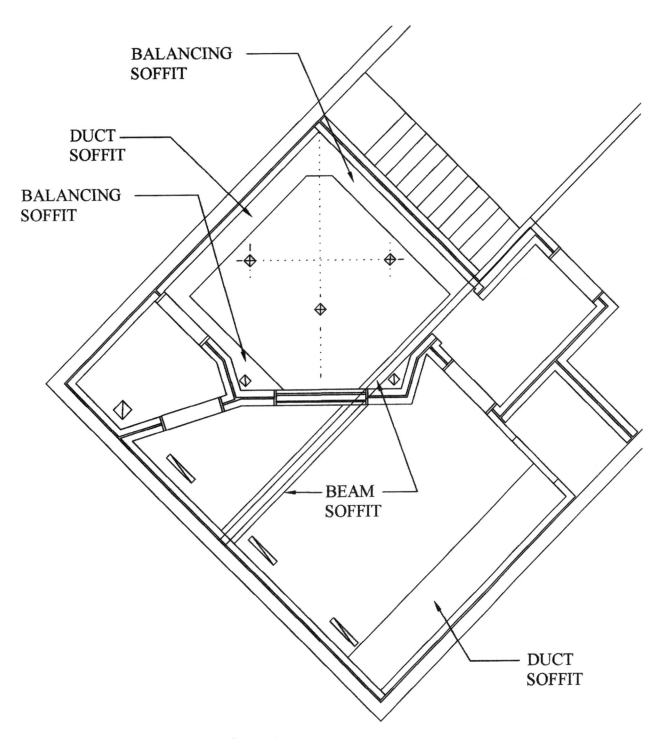

Figure 10.7 Rotated control room RCP.

You'll also see that there is a soffit in the corner of the control room to box in the exposed beam in the basement. Again, one is placed on the opposite wall to maintain symmetry.

That beam carries through the main room and creates a dead corner by the iso-booth entrance. The installation of the return air grille in that pocket will ensure air movement to that corner, helping to maintain even temperatures throughout the room.

The beam itself is a pain—a shame you couldn't lose it somehow—but it holds the house up so you live with it. Because it's below the floor joist, it makes it easy for your branch ducts to hide above the ceiling, and you can always mount strip lighting to it for a nice lighting effect.

Learn to use items like these to enhance your room rather than considering them to be just an eyesore. For ambiance, you can always set up a separate lighting circuit and place some backlighting behind this for the main room. Softly lit rooms can sometimes create the mood you need to get those creative juices flowing when you create your music. Lighting hidden behind beams creates great shadow effects.

By the way, that particular beam had a column located right next to the window into your control room. Your options are to leave it, or you could add steel flitch plates to either side of the beam and remove it to create an open view of the room.

If you opt for the flitch plate (which this design indicates, you can tell that because you don't see a column in the plan view), then make certain you verify with a structural engineer that the column remaining can take the added load because you might have to beef it up if it can't. Also, if you can't get the steel plates into the beam pocket at the foundation, you'll have to verify that the beam itself can take the added load at that location. It's possible that the shear load (the cutting action in a vertical direction at the foundation wall itself) will require an interior column to carry the added load.

Developing the Details

OK, so you know the design you want to use, you know the room layout, and you have a reflected ceiling plan. What's the next step?

The next step is to develop some elevations and sections through the rooms at various locations to determine exactly how you are going to build it. Figure 10.8 shows you where you would want to do those in this case.

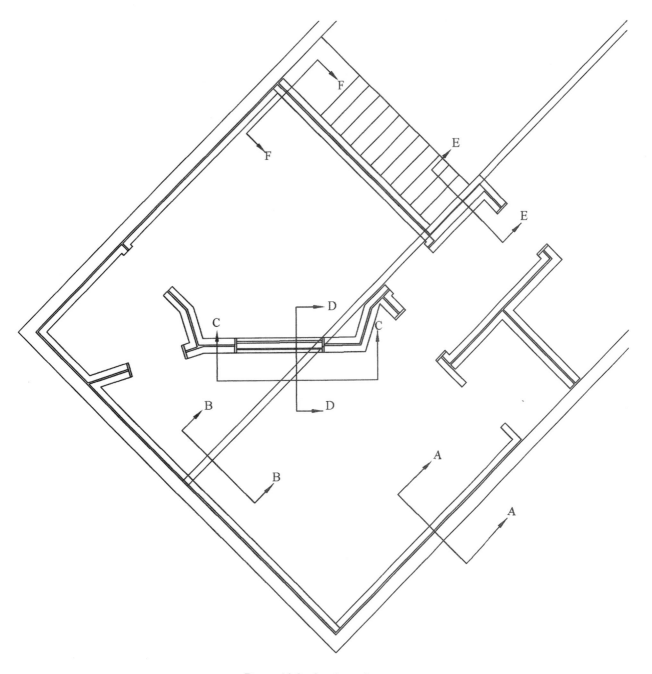

Figure 10.8 Sections directory.

These particular locations were chosen for the following reasons:

Location:

- AA—Where you have duct below the ceiling, this will develop a typical soffit detail.
- BB—Section through beam in room. You need to deal with this while maintaining isolation.

- ▶ CC—Framing out the window opening, while maintaining isolation from the beam (and the floor above) is going to be tricky. This will help you see that from this angle.
- ▶ DD—A second view of that same condition with the window/beam; this time splitting the beam.
- ▶ EE—Maintaining isolation from the structure above, while dealing with the carrying beam at the stairwell.
- ▶ FF—This view will develop a typical detail for the areas where duct runs above the ceiling.

You can develop as many of these details for particular areas as you want. Obviously, the more you develop for each different condition, the better off you will be when you begin construction.

These plans and details will also come in handy when you go to the building department in your town to pull a permit for the work. Demonstrating to the building official that you have carefully thought your way through this, and providing him with this level of detail, will go a long way toward helping him understand construction that he's probably not very familiar with, which will help you obtain that permit much more smoothly.

You also have to (based on the existing conditions) make up your mind as to how you want to proceed with construction. Is it going to be a true room within a room—with an independent ceiling and an elevated isolating concrete slab? Or is it going to be walls and ceiling only? Or will it be a hybrid of techniques?

Some of your decisions might be driven by building codes. For example, if you have 9' ceilings, you can afford quite a bit of vertical construction, while still maintaining the 7' minimum ceiling height proscribed by most building codes. If, on the other hand, you are pressed (vertically) for space, you won't have the same options.

In the case of this particular basement, the existing height to bottom of the floor joist is 7' 6" above the existing concrete slab. For this reason, to maximize ceiling height, the ceiling will be constructed using RC-2 and three layers of $5/8$" gypsum board. That would drop your ceiling height to 7' 3 $5/8$" ($1/2$" for the RC-2 + 1 $7/8$" for the three layers of gypsum).

The existing floor joists are 2x10 Hem/Fir #2 construction grade, installed at 16" centers. Out-to-out bearing is 11' 4 $1/4$", there is $3/4$ T&G decking above, with $3/4$" oak flooring above that over the live room, and carpet with pad over the control room/iso-booth. The total existing dead load is 7.68psf above the live room. Because this is the larger load of the two, you will concern yourself with not exceeding the allowable load for this area.

The space above the studio is split between living and dining rooms; thus, the required live loads are 40psf.

If you add five layers of ⅝" drywall to the structure (three on the ceiling—two above at the bottom of the plywood deck), with RC-2 on 24" centers—the total added dead load would be 13.08psf, for a total dead load of 20.76psf.

Calculating for a total dead load of 20psf, and increasing the live load to 50psf (for safety's sake) the 2x10 Hem/Fir is capable of a Maximum Horizontal Span of 12' 10" with a minimum bearing length of 0.98" required at each end of the member.

If you add one more layer of drywall, for a total of six layers, and add 10 more pounds to the total load calculation, then the joist would still be safe for a span of 12' 0". So your design will be three layers tight to the deck and three layers below RC-2 mounted to the joist. Your wall construction will be 2x4 at 24" centers, double wall, with three layers of drywall on each side, and standard R-13 insulation in each wall cavity.

Let's take a moment here to understand why you made the decisions you just did. After all, you're running through this fairly quickly, so you need to take a deep breath and make certain your choices are wise. You looked earlier at what creates isolation, and it's all about mass with an air spring. Or at least up until the point where flanking takes over. If you need a refresher on this, go back to Chapter 4, "Floor, Wall, and Ceiling Construction Details," and read it through again.

The ceiling/floor assembly will give you pretty good isolation; however, you are very limited in how much weight you can add to this, and you will find that the floor joists themselves are the weak link. This is because they (the floor joists) will penetrate the three layers of drywall and touch the ¾" decking, which is never an optimal design, although the drywall does help dramatically. For example, the center frequency of this deck will be around 19.3Hz, which will help keep your low-frequency sounds from penetrating the deck.

Another weak point here is flanking through the foundation into the structure above. Not only are you going to excite the foundation with transmissions going through the wall you construct adjacent to it, but there will also be transmissions through the slab that those new walls will rest on (basement slabs generally sit directly on the footing and touch the foundation walls).

You are also going to have some flanking (again through the slab) from room to room within the studio itself.

All of these items (combined) will define the maximum level of isolation you can achieve with your build-out, although a properly designed space will usually reach greater levels of isolation room to room in the studio than you could probably achieve between the basement and the space above. So everything begins with determining the best design for the weakest point.

As far as the deck goes, if you can afford another layer of drywall (assuming the structure can carry it), you could always add another layer of drywall to the assembly at the bottom of the ceiling. However, until you determine exactly what you have accomplished at this point in time (with the three layers we're already speaking about), you are better off waiting until the bottom three layers are up and then run a few sound tests to determine whether you need more than you have. No sense in throwing a fourth layer into the mix if you aren't going to need it.

Now that you understand the construction parameters, you can begin to develop elevations and sections. Figure 10.9 is a developed detail through Section AA. When developing wall sections, one of your biggest concerns has to be how to handle fire-stopping at the top of the wall assemblies. The building official will be looking for this for a very good reason—you would not want a fire that began inside of a wall cavity to be able to spread over your ceilings throughout the various rooms. You always want to compartmentalize your construction to keep a fire from spreading until you can either respond to it or escape from the fire.

Draw your existing foundation/footing slab and a section through the floor assembly, adding the interior wall and fireblocking, as shown. Note that the ceiling assembly (in this location) is going to end at the inside face of the duct chase. This places the HVAC plenum below the floor joist, located in the cavity between the two isolation assemblies. The chase itself is hung from the RC-2 to maintain isolation from the existing structure.

In order to brace the wall frame, it's important to use isolation clips attached to the structure in some manner. Although the attachment to the resilient channel is an effective means, in this case, it would interfere with the required fireblocking at the top of the wall assembly, so it's easier to attach to the foundation wall as indicated.

Note the gap between the ceiling and the firestop. Although you could place the framing tight to the drywall and achieve the same firestop with just a simple caulk joint (using fire caulk), this would create a short in the isolation system—a direct connection of the two structures. So always design to achieve the two separate goals, which are to stop the spread of fire and never directly connect the two structural elements.

In the case of the beam itself (in the tracking room where it's exposed), it's easy to maintain isolation by utilizing resilient channel from which to hang a soffit. Just make certain to keep the frames free from touching the beam as shown. A half-inch of clearance will be more than adequate. You can see this in Figure 10.10

You don't need any structure crossing the bottom of the beam to tie the two sides together because the drywall will serve that purpose. It's better to install the drywall on the bottom of the assembly before the sides, which gives you a chance to verify that the drywall locked the two sides into position without them touching the beam.

These types of frames (floating frames supporting a drywall load) are better when constructed with framing screws rather than nails because the screws have a greater pull-out load capacity than nails do, and you want this to stay up there for a long time without loosening up. Vibration of loosened members, years

Chapter 10 Putting It All Together

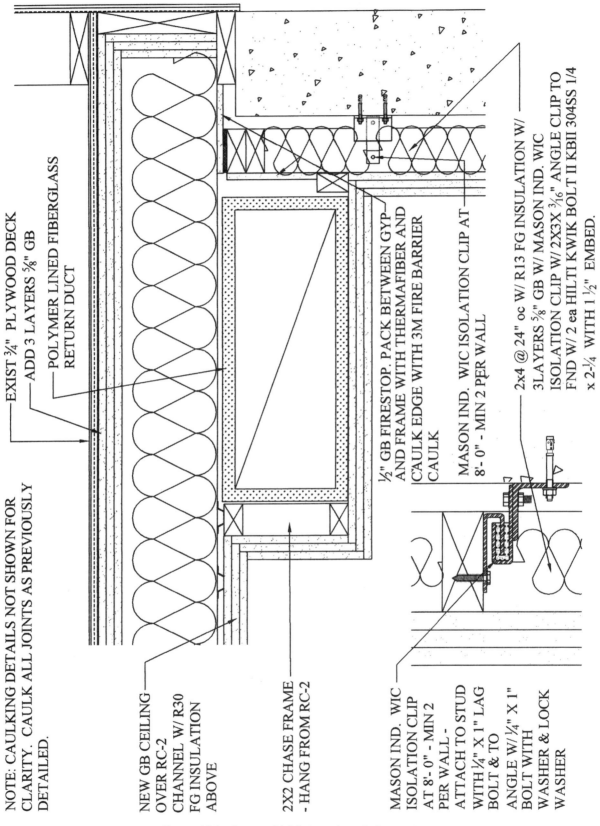

Figure 10.9 Section "AA" duct chase below ceiling.

down the road, could cause noises you do not want to hear in either your rooms or your recordings. So take the extra time here to screw the top and bottom plates to your mini-studs when constructing the side frames.

Figure 10.11 shows you how to frame out the window opening into the control room, while maintaining isolation from the beam. Details like this will help you see the difficulties involved with that area. In the case of this window opening, you have the building's carrying beam crossing over the window opening. This not only affects the window height, but also how you go about framing the window, while maintaining a header assembly that will support itself over the years.

In this particular case, it's easiest if you just let the window header (a single 2x4 on the flat) run past the jack studs (on the right-hand side of the window) to the next stud layout as shown. Frame the remainder of the window opening and then create a truss by installing ⅝" plywood attached to the frame.

I know that it might seem to you that the truss will be ineffective with the big chunk of material you have to cut out to accommodate the beam, but that isn't true. First, the header assembly is not really load bearing. (You'll understand that better when you see the next section through that wall.) Second, the cantilever effect created by carrying the header out to the next stud bay goes a long way toward stiffening this up.

Figure 10.12 is another view of that same condition, this time looking through the centerline of the two walls. Note how the two top plates are separated from one another and how the inside wall of the control room never makes contact with the ceiling.

You have effectively designed a slip joint assembly, allowing the beam (and subsequently the floor above) to deflect under various loading conditions without imparting any of that load to the header assembly below. This is why the comment was made previously that the header is not really load bearing. It only has to support its own weight. All of this work is performed on the main room side of the wall assembly, with the inner wall bracing being provided by the WIC isolator clips attached to that wall.

Note that for this slip joint to work properly, no fasteners are installed connecting the plywood to the uppermost plate. This plate must be free for movement. It is very important that the beam/floor assembly—under loading—does not impart any of that load to the window frame.

This will be a typical detail for all of your interior walls, minus the plywood header. One top plate attaches to the RC, and the other wall attaches to the first one through the use of an iso-clip. The only difference is that the remaining carrying walls do not require separated top plates like you used here. They can touch. All of the walls will require the firestop above the lower wall, effectively closing off the gap between the two walls into the ceiling cavity.

Chapter 10 Putting It All Together

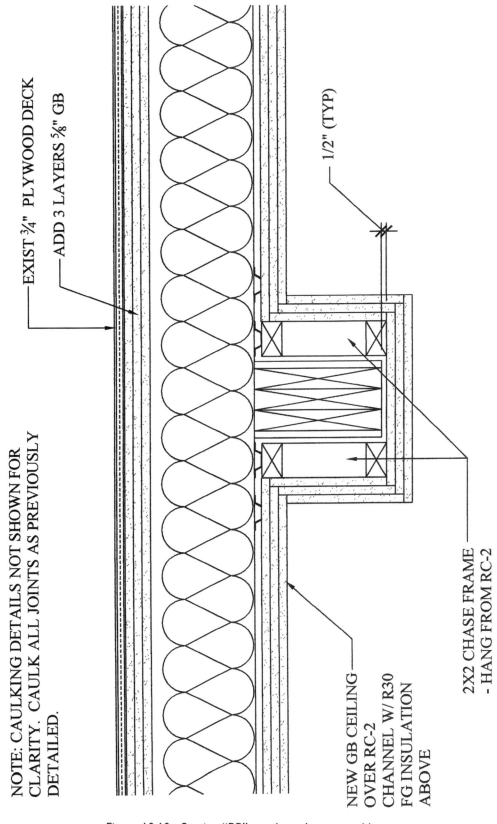

Figure 10.10 Section "BB" cut through structural beam.

261

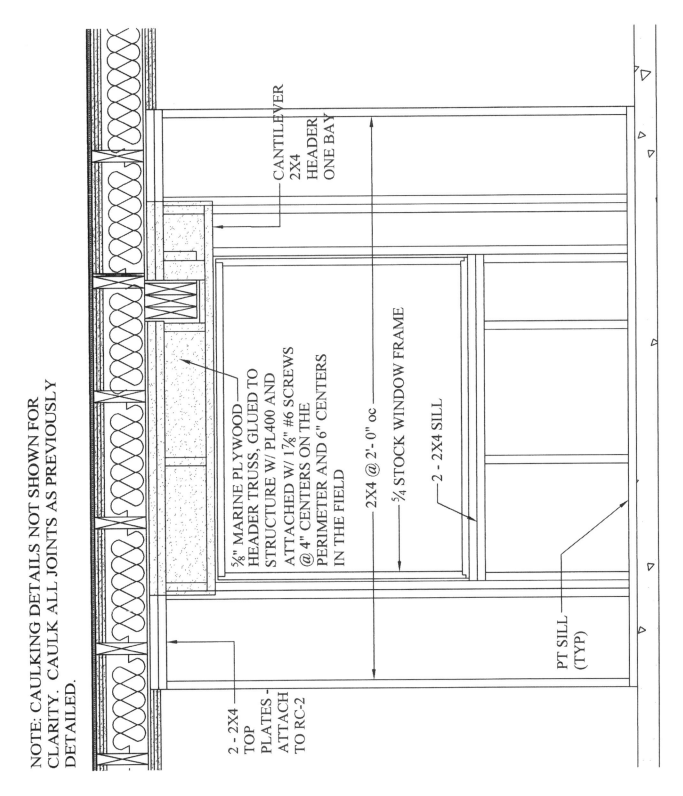

Figure 10.11 Elevation "CC" control room window at beam.

A comment regarding that. By now you're probably wondering why you don't just carry the ceiling throughout the whole area to begin with and do away with all of these firestops you are going to have to put in. It would be a whole lot easier, right?

The answer is that it would be a whole lot easier, but it would not afford you the same isolation. Connecting the ceilings on both sides of the walls would create another pathway for sound to travel from room to room, so you would decrease your isolation by doing it in that manner. Remember that maximum isolation requires a disconnecting means between rooms. The RC-2 does that from room to room, as well as from downstairs to upstairs. The details shown are nice, clean, neat, and they provide the isolation you want to achieve.

Isolating the air lock from the house structure is as important as dealing with it anywhere else in your studio. Remember that air locks actually give you a slight decrease in wall performance, based on the fact that they create four-leaf systems. This is another item you examined in Chapter 4. However, this is one of those, "Oh well, we live with what we have to live with" deals. Remember, what you can't change, you live with. But just because we have to live with it doesn't mean that you can get sloppy with the details. Plus, the fact that the four-leaf system degrades your isolation makes it even more important.

Take a look at Figure 10.13. It's a cut through the air lock wall adjoining the stairwell.

Note that the wall below the beam is inset 5/8" and attaches directly to the beam. No problem with this direct contact at that location. The outer sheets of drywall lap the beam edge. You should run a bead or two of acoustic sealant behind the drywall to seal it to the beam. There needs to be a way to tie the WIC connector to the other wall (which in this case is too far from the outer wall to make a direct connection between the two). This can be accomplished through the use of a support made from two pieces of 3/4" plywood glued and screwed together. Clip these to the beam using four angle brackets and nail it to the backer you see above the firestop. The firestop should be tight fitting to this support member when installed.

Once again, the inner wall is not tied directly to the existing structure. It's 1/2" below the firestop with the typical detail of compacted 4psf rockwool and a fire-caulk seal.

Focus on the dropped ceiling in this space. It's dropped due to the HVAC supply duct that runs through the air lock. The first layer of drywall on the wall passes above this ceiling to complete the firestop for the wall cavity. The second and third layers interlock with the drywall on the ceiling, using the standard backer rod and acoustic caulk detail (which is not shown in this case, just for the sake of clarity). A simple block can be attached (to the side wall) as shown for ceiling connections, although you still have to hang the mid ceiling supports from the RC-2, as indicated.

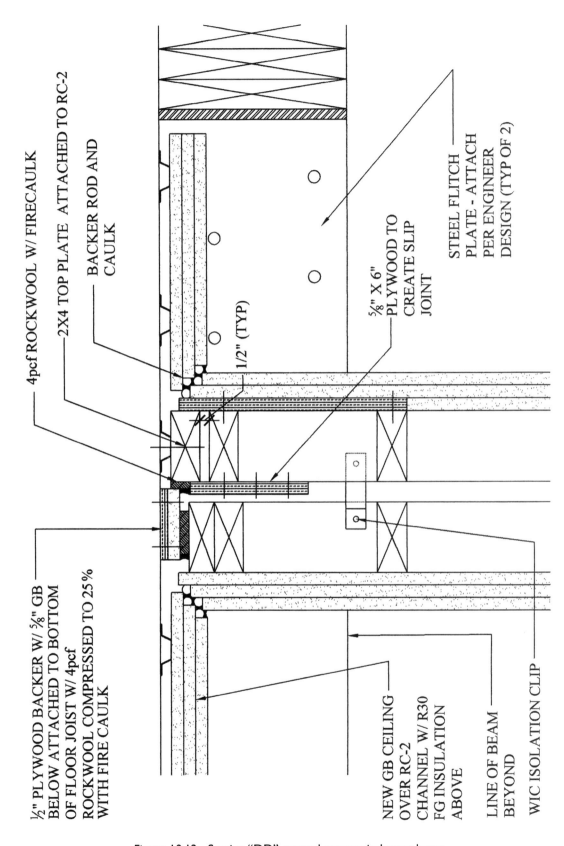

Figure 10.12 Section "DD" control room window at beam.

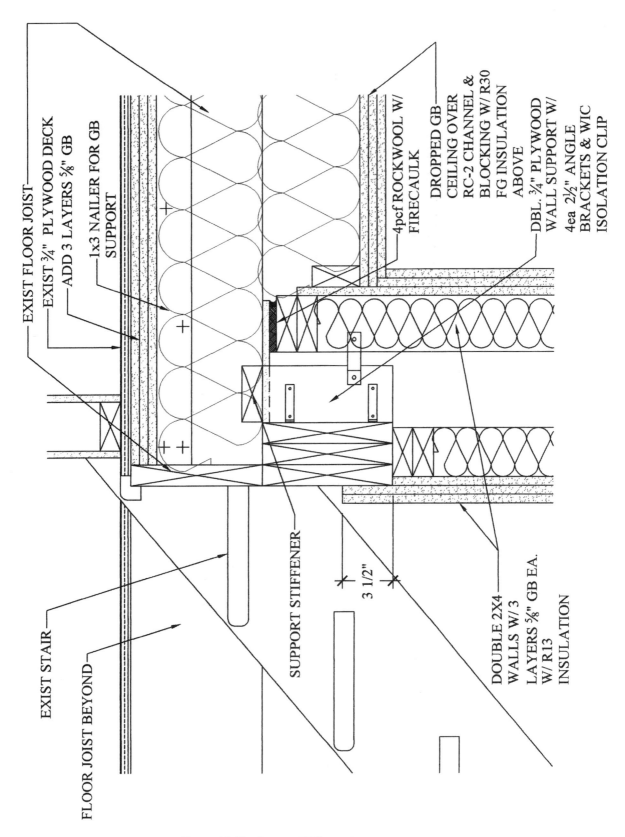

Figure 10.13 Section "EE" cut through beam at stair.

Figure 10.14 is a smaller scale detail intended to give you a better perspective of the entire air lock. Note that the opposite wall is basically a mirror image of the one on the left, as is the wall/ceiling assembly within the closet (where the HVAC return air duct is located). The ceiling is located just 1" above the top of the door in this area to maximize the available space for the HVAC duct.

Finally, Figure 10.15 is a section through the area where duct runs above the ceiling. This is a good detail to develop because it also has to deal with the end condition for RC-2 channel, something you haven't looked at previously.

Stop the channel in this condition before the inside face of the wall ($1/2$" to $3/4$" is acceptable). Drop the top of the inner wall $1/2$" below the bottom of the joist and install the drywall $1/2$" above the bottom of the joist, using blocking as shown (attached to the existing floor joist). Pack the 1' joint between the two surfaces solid with mineral wool; then just caulk the joint between the top of the wall and the bottom of the joist with fire caulk. This is much cleaner to install than if you ran the RC-2 past the wall and used it for attachment (in lieu of the WIC clip). The reason for this is that it's much easier to pack a wide open joint than to pack underneath (and around) the RC-2.

Caulk details for the drywall joints are the typical interlocked drywall with backer rod and acoustic caulk. You have some examples of this in Figure 10.12.

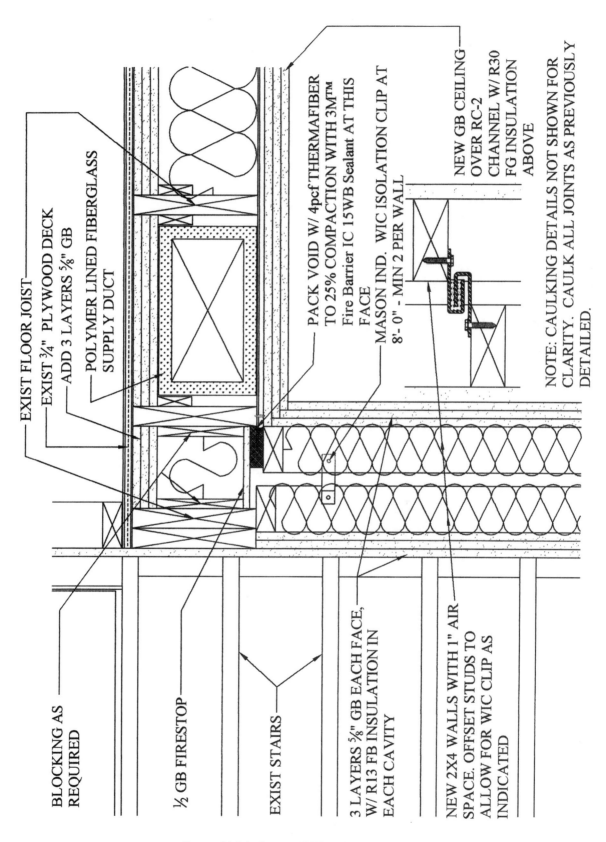

Figure 10.14 Section "EE" wider view.

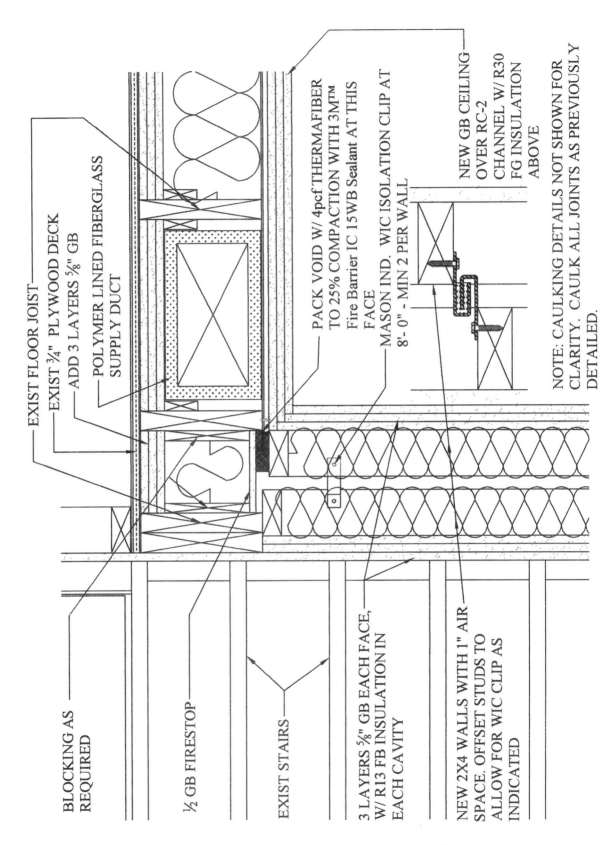

Figure 10.15 Section "FF" cut through duct chase above ceiling.

A Few Different Case Scenarios for Your Perusal

Not everyone will be building in a basement. Some of you might build in a garage, others will construct in attic spaces, and yet others will have to make use of rooms inside of your home. So we will take some time here to look at options you have when faced with construction in those areas.

From a purely design point of view, the concepts of room shapes stay the same as when working in a basement, so the focus here will be on how to isolate these spaces from those surrounding them.

An Inexpensive but Effective Decoupling Floor System

Since the first printing of this book, I was faced with the challenge of creating a decoupling design for a studio located on the third floor in a building that could not support an elevated concrete slab.

This was a pro studio and had to be capable of operating 24 hours a day, so it was "get creative" time.

The solution was a lightweight assembly utilizing 3psf rigid fiberglass along with a couple of layers of plywood.

It was my first time using this system, and I could find no data to prove that it would work. The owners were not going to pay for the construction of a mock-up of their floor assembly and this system in order to have it tested, nor were they going to pay me to travel halfway around the world in order to do a mock-up on site, so this one was going to have to be based strictly on intuition and construction experience. The proof that this worked came in the form of a decision that the owner made during construction.

The concept here was to install the walls first, including the drywall finishes, after which we placed 1 $1/2$" to 2" of rigid insulation over the entire floor of the room. This was then covered with 2 layers of $1/2$" plywood with overlapped joints; we screwed the plywood sheets together using 1 $3/4$" screws. If you want (as we did there) to try to get a little more isolation out of this, you can use Green Glue between the plywood sheets. Figure 10.16 is a section cut through the bottom of a wall and floor detailing how this should be arranged. Pay close attention to the caulking details here, as well as the general concept.

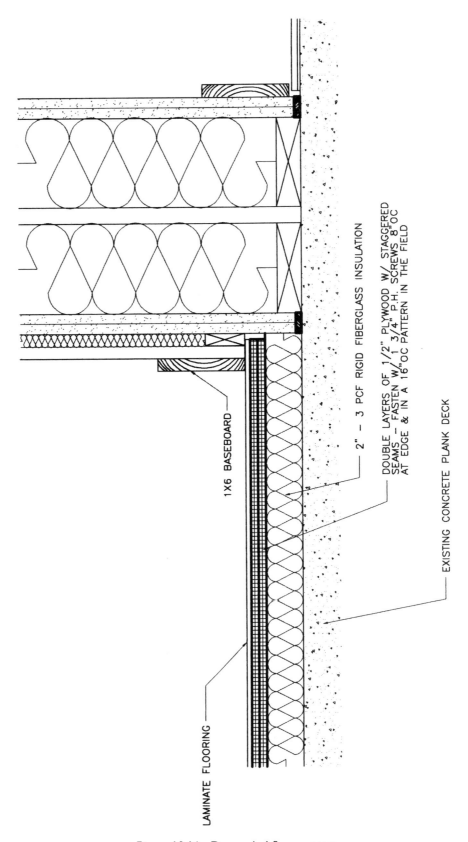

Figure 10.16 Decoupled floor system.

Oh, I almost forgot—to clarify the prior comment about the owner of this facility making a decision that proved the worth of this system, here is what happened. In the Control Room, the owner decided to gain 2 ½" of headroom and thus didn't install this system there. After completion of the room, they could hear (in the control room) air-handler fans in the space below when they were running. This was due to flanking taking place through the concrete deck separating the two spaces, to which the air handler was directly connected. They ended up having to invest in the installation of isolation hangers for that air-handler unit, which solved the problem.

The other proof (of course) is that even with full bands working in the live room, there was no transmission to the offices below.

Since then, I have successfully used this system on other projects, including a professional-level mixing room (with an iso-booth) in the home of a client located in Brooklyn, NY. This client wanted a high-quality room where he could take work home, so he could get some away time from the studio without having to miss deadlines.

He was insistent that the studio be located on the first floor facing the street, and that the existing window was to remain so he could enjoy the natural light. This was also complicated by the fact that he was in a unit that was flanked by living units on both sides.

The physical separation for these units was constructed using multi-wythe brick walls, and the floor joists were supported by those same party walls. A *wythe* is a single brick in depth, and a brick veneer over the face of a wall is single wythe in nature. Multiple wythes are adjacent brick structures tied together by either spanning between the walls with bricks, or through the use of metal ties made specifically for that purpose.

Again, the use of this system in the floor assembly helped to ensure isolation between the various units.

Beefing Up Garage (or Other Exterior) Wall Assemblies

One of the challenges when building inside of an existing garage is the lack of mass on the exterior walls. You can get around this (without removing the siding on the building and incurring the costs associated with doing anything that drastic) by working *within* the wall system itself.

If the garage already has drywall installed in it, you can make use of this by carefully cutting the drywall flush to the inside edges of the studs and plates using either a reciprocal or drywall saw.

Before doing this, make certain to turn off any and all electrical circuits running to the garage. It is just as easy to cut through electrical wires as it is through drywall. Do not assume that because a wall has no outlets that there are no electrical wires running through it. Electricity is extremely dangerous, so you should always proceed with caution.

Use these pieces (or new drywall if your garage was not finished inside), fitting them so they have a $3/8"$ gap between them and the framing. (This is for the purpose of creating a caulk joint to ensure an airtight seal.) I would recommend that you place at least two layers of drywall or more if you are in a particularly noisy area.

Attaching the drywall is dependent on what you have for building sheathing. If your garage is older and has board-type sheathing, you can directly attach to that using screws. If it's $1/2"$ plywood sheathing, you can use the method outlined for decks and use wood furring to lock it into place.

Figure 10.17 is a section through a garage wall detailing this method of beefing up that wall. Note the use of backer rod and caulk.

Continuation Lines

Those funny little lines that you see in this picture (where the center of the wall disappears) are called "continuation lines." They indicate that the section they connect to continues in the direction of the continuation line (which, in this case, would be until all of the wall lines join). They are an easy method of detailing the top or bottom of a wall assembly without having to draw the entire wall, which then allows for larger scale details than what would be possible with the full height wall assembly. In this particular instance, it is easy to see that the wall construction (between the continuation lines) is exactly the same as what you see up to those points, top and bottom. The sheathing, stud, insulation, and drywall all continue unbroken until they form the complete wall assembly. Nothing is different in the middle.

When things *are* different in the middle, for example, buildings with numerous floors and different details at each floor level, (exterior trims for instance), you can use them to break out each different detail so that everyone knows that all of those details exist on the same side of the building, while calling out the elevation so that they know exactly where the detail occurs.

They can also be used in plan view drawings with the same effectiveness.

Figure 10.18 is a picture of this application in place.

Chapter 10 Putting It All Together

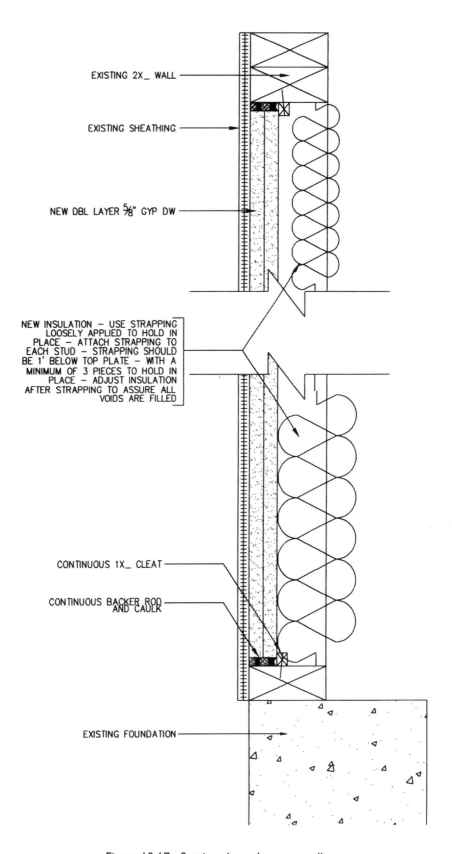

Figure 10.17 Section through garage wall.

273

Figure 10.18 Exterior wall isolation.

Attics and Other Room Spaces Inside of Your Home

Here you have your work cut out for you. Building a space over other living areas and obtaining any sort of real isolation is one heck of a job. If you're lucky, the attic or room is over a garage, which makes it a little easier. However, even then, overcoming flanking is a real challenge.

There are things, however, that you can do to maximize isolation in these cases, even if you can't construct a room within a room.

To begin, this would be a case where I would strongly advocate the use of isolation clips—the better the clip, the better the isolation will be.

Next, you should have a structural engineer examine your floor assembly to determine exactly how much weight it can carry because you are going to need to add mass to this to increase isolation. If you can do anything with the ceiling below, try to add RISC-clips to decouple insulation in the cavity and mass to that surface. If you cannot touch that surface for one reason or another, then you will add all of your mass and decouple the floor from above.

Figure 10.19 is a section through the floor of a building using some of these techniques. The mass in this case was achieved by using lightweight gypsum concrete (about 7psf of mass) on top of a 3/4" T&G (tongue and groove) deck. Note again the use of rigid insulation with plywood above to create a decoupled floor assembly. Further isolation could be achieved by floating the wall above the existing deck, assuming the structure could support the point loads being imposed on it. This will be discussed when we examine Figure 10.20.

Chapter 10 Putting It All Together

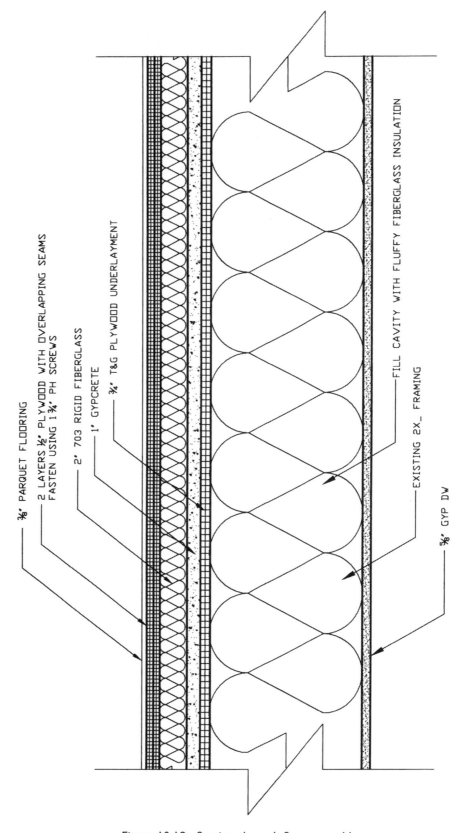

Figure 10.19 Section through floor assembly.

Point Loads

Typically, loads from a wall are distributed evenly to the structure below. The studs bear on a bottom plate, which helps to spread this load beyond the point of the individual studs themselves.

A point load is created when you take a section of the load and carry it by a single point instead of spreading the load throughout the length of the assembly.

I am quite certain everyone reading this book has seen columns in a basement, carrying a porch roof, or instances similar to these. The columns are perfect examples of point loads carrying structure above. The beams they carry are equally good examples of methods of spreading the loads above from wall to column or column to column.

Floating Your Walls.

In Chapter 4, you viewed Mason Industries' "ND Isolators," which can be used (quite effectively) to create a floating wall assembly that is decoupled from the floor it rests on. Here you will examine how to use those products effectively.

Floating Assemblies

The principles applied here using ND Isolators are the same, regardless of the type of isolator you use. Whether it is the ND, a spring isolator, or just neoprene pads, the concept never changes. You must first determine where you want the isolators placed, and then what the load is for each of those isolators.

One reason I prefer to use the ND Isolator is because it is tailor-made to provide an attaching point to the structures above and below it without creating any sort of flanking path. It is also more cost-effective (in this case) than a spring-type isolator.

It takes a little bit of time (and some simple math) to determine which model of this product to use because they all have different load requirements in order to work properly. However, once you have the concept down, determining this for your own build is not really all that difficult.

For the purposes of this exercise, we will look at a wall that will have a super door installed as a part of the finished product. This wall will be 20' 0" in length, made using 2x6 construction, with the studs installed 24" on center, and with two layers of ⅝" drywall applied to the face of the studs.

It will have a double 2x6 top plate with a single 2x6 bottom plate, and will have 2x8 members let into the studs at both the top and bottom of the wall to distribute the wall loads evenly to the isolators below. A 2x6 sub-sill will be fastened to the deck prior to installation of the isolator, to which the isolator will be attached.

You will begin by creating a simple elevational sketch of the wall frame along with a simple plan view. In Figure 10.20, you can see both an elevation/section and a plan view of a typical wall using these isolators. This is the level of detail you really need to develop in order to work this out and keep things straight during construction

The plan view helps you picture the stud configuration for the wall corners; this is an important piece of your puzzle because without it there is no way for you to visually create the actual corner condition.

Take the time to develop this level of detail as you work your way through your studio design. It is well worth the effort for a whole variety of reasons, even if you aren't using this isolator. It gives you a very accurate hard count of the materials going into your build, which is worth its weight when it comes down to creating a truly accurate estimate of the costs you are going to incur.

After you have the sketches finished, you can begin your analysis. Let's take a look at how to calculate the right isolator for each location.

You begin by determining what materials you are going to use in your construction and then verifying those material weights. Lumber weights vary based on species. You can find the data for your lumber on the Internet with a simple search, or you can go to the lumberyard for the information.

Douglas Fir lumber weighs 35 pounds per cubic foot , and is what I would typically specify for the framing in any wood-framed project. So we will use that material in this exercise.

The material weights for various Doug Fir members are as follows:

▶ 2x4 weight: 1.28 pounds per linear foot

▶ 2x6 weight: 2.00 pounds per linear foot

▶ 2x8 weight: 2.64 pounds per linear foot

▶ 2x10 weight: 3.37 pounds per linear foot

▶ 2x12 weight: 4.10 pounds per linear foot

You have a wall constructed using 2x6 lumber, with three plates and a 7' 11 1/2" wall height. The studs are short of the wall height by 4 1/2", (the thickness of the three plates combined) so the studs are 7' 7" in length (7.58').

You also have a 2x8 at the top and bottom of the wall, "let into" the studs (which means notching the studs to allow the member to be flush with the wall when attached), and two layers of 5/8" drywall, which weighs about 5.2psf.

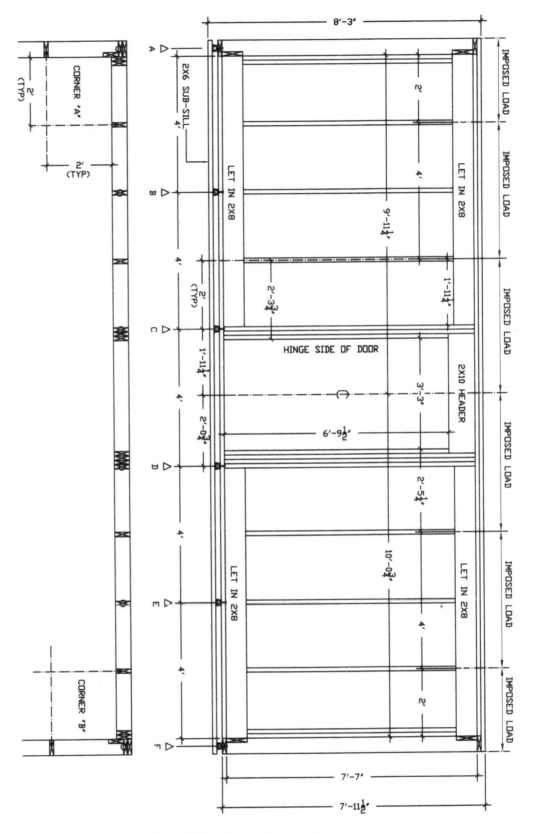

Figure 10.20 Wall design elevation and plan view.

You will calculate assuming the placement of isolating supports on 4' 0" centers.

When you calculate for these types of loads, each support carries exactly ½ of the load between it and the next support on either side. So, working from left to right, the first support is 4' (to the center of the stud) from the inside face of the corner, and carries the load from the center of the stud on its left [located 2' 0" (centerline from the corner) to the center of the stud 2' 0" to its immediate right (6' 0" centerline from the corner].

The span that creates the imposed load for each isolator is indicated above the wall assembly.

Understanding the Imposed Loads

In the vast majority of cases, imposed loads are evenly spread between supports. What this means is that the weight distribution between two supports will be exactly one-half of the total load. So with supports located on 4' centers, each support will carry exactly 4' of the supported structure, 2' on either side. You can see that this is the case with all of the supports in Figure 10.20, except for the supports at the corners.

In this case, the two side walls are sitting directly on the supports, and thus none of that load is being transferred to the wall constructed between them. The load imposed on the supports for the back wall (the wall running left to right in the bottom of the figure) begins at the inside face of the corner.

If the back wall (in this case) was 20' 11" long and ran past the two end walls, then the load would be calculated from the center of the corner support to the halfway point in between the two supports.

We will examine corner calculations at the end of this exercise.

One caution here would be that this would not be the case if you began imposing loads that were unbalanced. For example, the load of a simple floor joist applied to a pair of walls will always be equally spread between the two walls; however, in case of an additional load carried by that joist that was off-center, the load would not be evenly applied to both ends of those same joists. As an added load gets closer to one wall, a greater percentage of its load will be carried by that wall.

An example of this would be an air handler installed for HVAC that was carried by the ceiling joists above.

Another example would be carrying a load where the end of the wall did not have a support (in other words, the end of the wall was a cantilever that extended beyond the last support). In this case, the unbalanced load may well decrease the actual load on the second support. Picture a seesaw with a person on one side weighing 100 pounds and a person on the other side weighing 115 pounds. If they just sit and let everything settle, the lighter person will be higher in the air than the heavier person. In this case, you could either calculate the actual load at the second support, or you could increase the distance between the two supports to regain a balanced load.

In either case, the calculations become a bit complicated, and your best bet would be to have a structural engineer calculate the imposed loads for you prior to choosing your isolator.

On to the actual calculations:

The load at point "B" would be $1/2$ of the stud weight at the 2' point, the stud at the 4' point, and $1/2$ of the stud at the 6' point, so a total the weight of 2 studs, 33s.f. of drywall, 8l.f. of 2x8 (4' each at the top and bottom of the wall), and 12 linear feet of plates.

The stud load is calculated as: 7.58l.f. * 2.0plf * 2 studs for a total of 30.32lb.

The 2x8 load is: 8l.f. * 2.64plf for a total of 21.12lb.

The drywall load is: 33s.f. * 5.2psf for a total of 171.60lb.

The plate load is 12l.f. * 2.0plf for a total of 24lb.

Thus, the total pin load for that isolator is 247.04lb.

The load chart below provides the information you will use to select the proper ND Isolator to suit your needs. Note that these are color coded so that they are easy to identify in the field during installation. The "B" size White and the "C" sizes Green and Red all fit the bill. The smaller sizes are less expensive than the larger ones, so White "B" would be a good choice for this location.

TYPE ND RATINGS

Size (Color Mark)		Duro-meter	Rated Capacity Range		Max Rated Deflection	
			(lbs)	(kgs)	(in)	(mm)
ND A	Black	30	15 - 45	7 - 20		
ND A	Green	40	30 - 75	13 - 34	0.35	9
ND A	Red	50	60 - 125	27 - 57		
ND B	Black	30	50 - 100	23 - 45		
ND B	Green	40	75 - 150	34 - 68		
ND B	Red	50	110 - 235	50 - 107	0.40	10
ND B	White	60	180 - 380	82 - 172		
ND B	Yellow	70	520 - 600	136 - 272		
ND C	Green	40	140 - 260	54 - 118		
ND C	Red	50	200 - 400	91 - 181	0.50	13
ND C	White	60	310 - 600	141 - 272		
ND C	Yellow	70	520 - 1000	236 - 454		
ND D	Yellow	70	481 - 953	481 - 953	0.50	13
ND DS	Yellow	70	998 - 1950	998 - 1950	0.50	13

The bay to the left of the door opening, and one-half of the opening itself will be carried by the next isolator. This isolator will also be carrying the entire weight of the door assembly.

As you can see, the wall load is 2' 3 3/4" to the edge of the door frame opening from the center line of the stud that ends the imposed load from the first isolator we just sized. There are three and one-half studs plus 1/2 of the door header (which is a double header). The jack stud supporting the header is 6' 9 1/2" (6.79').

The stud load is 7.58l.f. * 2.0plf * 2.5 studs for a total of 37.9lb.

The jack stud adds 6.79l.f.* 2.0plf for a total of 13.58lb.

The "let in" 2x8s are 2.33l.f. * 2 each * 2.64plf for a total of 12.3lb.

One-half of the header is 1' 9" (1.75') * 2 each * 3.37plf for a total of 11.8lb.

The drywall is the width to the door plus one-half of the door header—a total of 20.82s.f. * 5.2psf for a total of 108.26lb.

The top and bottom plates are only a total of 10' 2" (10.166l.f.) (because the bottom plates are cut flush to the inside of the jack stud) * 2plf for a total of 20.33lb.

The total of the loads above is 204.17lb.

And now comes the door. For this exercise, we will calculate the weight of a super door.

Saving Yourself Some Time

You could go through the exercise here of calculating the entire weight of the assembly, hence the door as designed, hinges, door frame, closure, threshold, weather-stripping, etc. However, you can shorten the process considerably by just calculating the door itself and then checking the total load against the isolator that will carry it. If you are sitting somewhere around the middle of the capacity, and the spread is more than 50 or 60 pounds from the outer edge, then you can forgo calculating this down to the last screw and bolt.

If you are somewhat close to the capacity, then do the long calculation.

A good average weight to use for a 1 3/4" solid-core 3068 door is 90lb. (of course, if you have already purchased your door, you can use the exact weight), the lead is 8psf, and 3/4" plywood is about 60.8lb. per sheet.

A 3068 door is 20s.f. in area, so it is 90lb. (the door) + 20s.f. * 8psf (for the lead) + 60.8lb. (the plywood face) for a total of 310.8lb.

Now, using this plus the totals above (204.17lb.), we arrive at a point load of 514.97lb.

That weight would be supported just fine using a ND-B-Yellow, which has a range of 300–600lb. and allows enough freedom so that you don't have to calculate for each and every component of the door assembly.

The right side of the door frame is just 1 1/2" wider than the left, has the same studs and jacks, and weighs almost the same as the left (minus the door weight). So look first at what works with that and see if you really have to calculate for that extra 3" of drywall and 4 1/2" of plates that make up the difference.

The weight minus the door is 204.17lb., which would work just fine with the ND-B-Red. There is no reason to bother calculating the additional drywall or framing.

Finally, you have the wall corners, after which you can repeat the exercise for the remaining walls.

As you can see, the studs on the flanking walls are 2' 0" from the corner, so the drywall weight will be the same as it is for the first section we calculated—33s.f., thus 171.6lbs.

The "let in" 2x8s top and bottom are also the same as that section—21.12lb.

The top and bottom plates, however, have an extra 5 1/2" in the corner, for a total of 1' 4 1/2" of additional framing. Again, don't sweat the small stuff—1' 4" is fine—for a total of 13.33l.f. at 2.0plf, which is 26.66lb.

That leaves the studs themselves, a total of two half studs in the field of the walls and five in the corner itself. That makes six studs total so 7.58.lf. * 2.0plf * 6 each, which equals 90.96lb.

The total load at the corner is 310.34lb.

The ND-B-White is the right isolator for the corner.

When you are done with all of your calculations, you should mark up another sketch indicating exactly what goes where, along with the loads for each location, so that you won't have to figure it all over again when you finally begin your build.

Figure 10.21 below is an example of what the sketch should look like when you are finished.

If you take this much care with your design, then using this type of system will stop the energy in the body of this wall from finding a flanking path into the original structure, at least through the floor. You will still have to take the same level of care at the ceiling.

Make sure to keep copies of both of these sketches, so that you can provide them to your contractor (or use them yourself if you are doing the construction) when it comes time to build these assemblies.

One other piece that you should consider is the construction sequence that you use after the wall is constructed. Any isolator is going to settle as weight is added to it. The maximum amount of settlement indicated by the manufacturer is calculated based on full loading, but this does not mean that there

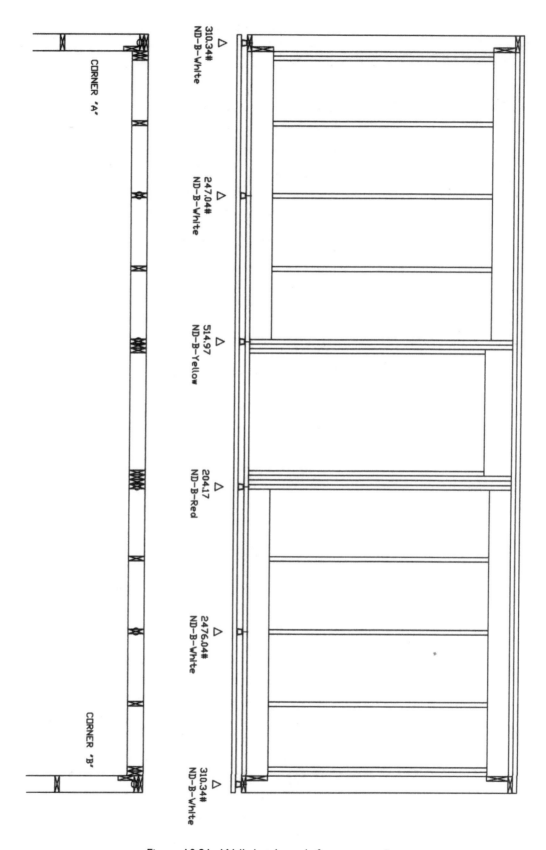

Figure 10.21 Wall sketch ready for construction.

is no settling taking place with a minimum load applied. There is also a certain amount of "creep" (additional settlement) that will take place over time due to fatigue of the isolator from long-term applied loads. You can contact the manufacturer with the actual loads, and they will let you know how much their isolator will settle with both imposed loads and fatigue.

This is an important piece of information because the space you leave between the drywall and the floor has to allow for this factor. The joint has to be constructed so that the drywall can settle, and the compression of the caulk joint does not cause a caulk failure when it does.

I would also suggest that you seriously consider installing your door before drywall if you use a super door design in your studio. A super door will be the largest contributor to the load of the isolator carrying it, and you are better off having that load on the isolator before installing drywall than after. You can always cover the door (and jamb) with protective paper to protect it from damage before you install drywall, tape, and wall finishes.

Dealing with Existing Roof Structures

If you are building your room in an attic or in a garage, you may well reach a point where you have to deal with the dreaded three-leaf system when it comes to the existing roof assembly. This is due to the fact that the building code (at least here in the U.S.) requires ventilation of the area directly below the roof deck. This could be through the use of properly sized gable louvers, or if the underside of the rafters is going to receive drywall, by continually venting each rafter bay.

There is no code-compliant way to avoid this.

I have listened to many people suggest that one way out of this is to add mass to the underside of the roof deck, insulate the rafter bays, and then add a vapor retarder to the underside of the rafters before installing drywall. While this may sound like a good suggestion, it isn't, for a variety of reasons. The most important reason is noted previously: *It is a building code violation.* And by that I mean, not the placement of the vapor retarder, but the elimination of the ventilation.

The building codes do not allow this as an alternative to the requirements for ventilating the space. It is just that simple.

You may think that the plastic sheeting you buy in rolls is a vapor barrier, but when all is said and done, it really isn't. Although vapor won't pass easily through the body of the sheeting itself, there are numerous penetrations through the barrier.

The staples you use to hold it in place until the drywall is installed, nails or screws through the drywall, joints between pieces of the sheeting, the conditions at the bottom of the wall— all of these are places where vapor can enter the rafter bay—and (upon reaching the roof deck itself) condense on the structure causing water damage, mold, etc.

There are pretty much two options that exist. You can deal with this on top of the roof, or you can live with a three-leaf system and construct the assembly with the intent of overcoming the issue.

Dealing with this outside of the building envelope would be the best option, and it is easy enough to incorporate into your work if you are constructing a new building. Simply add all the mass you want (within structural limitations, of course) and then insulate the top of the roof with insulating panels made for this purpose, add your shingles or other roofing material, and you now have a mass leaf that does not require ventilation below.

However, if you are working with an existing building, this may not be an option for you. If that is the case, then the following will lead you through some methods you can use to work your way around this.

Figure 10.22 is a section through a cathedral ceiling that details the conditions we are discussing. Note the use of "proper vent" to maintain the air space between the bottom of the roof deck and the insulation.

Saving Yourself Some Time

Proper vent is an inexpensive Styrofoam product made specifically for this purpose. It is made in sizes to fit between the rafter bays for both 16" and 24" oc framing. It has flanges on either side of the vent, plus a flat section about 1 1/2" wide in the center of the vent that makes contact with the roof deck so it can be trimmed to fit in bays that are not full width. This is much easier (and less expensive) than installing cleats in both sides of the bay and then installing plywood beneath the deck to create a vent space.

Install insulation in the bays and then install multiple layers of drywall directly to the underside of the rafters. When this is completed, install your inner room framing, leaving enough room between the bottom of the drywall above to allow for routing of fresh air and other ducting or electrical wiring as may be necessary.

You really do not want to build this as tight to the upper assembly as may be possible, because the larger the cavity, the more isolation you will have when all is said and done.

Once the framing is completed, install ducting and electrical lighting feeds and then insulate the ceiling bays and install multiple layers of drywall for the inner room assemblies as a whole. Don't forget to do this one layer at a time: first on the ceiling, then on each wall panel, so you can install backer rod and caulk for the entire perimeter of each panel assembly as you proceed. Maintaining an airtight assembly is critical to ensure maximum sound isolation.

One other thing you can do (that adds mass at the same time as granting you another benefit) is to install structural sheathing to the inside of the inner room to create a completely free-standing room. Doing this effectively creates a decoupled structure while giving you slightly more mass for isolation.

Figure 10.23 is a full building section through this room in the garage.

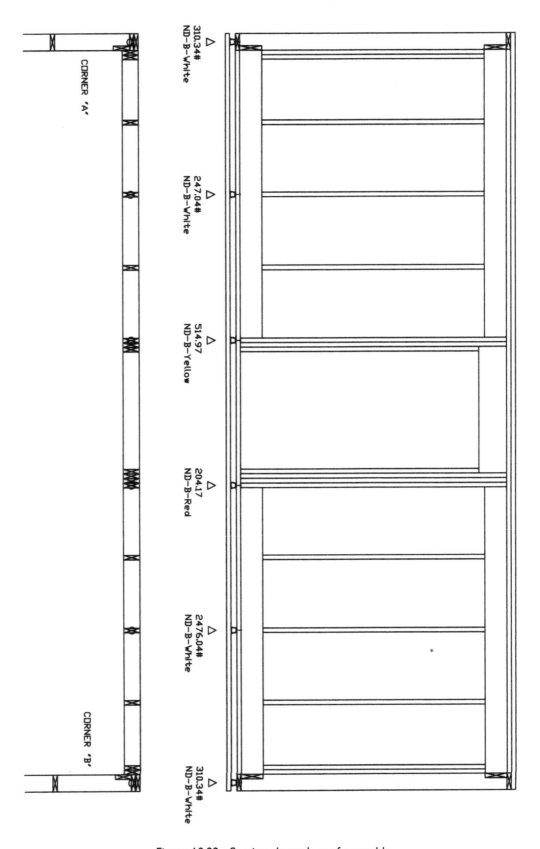

Figure 10.22 Section through roof assembly.

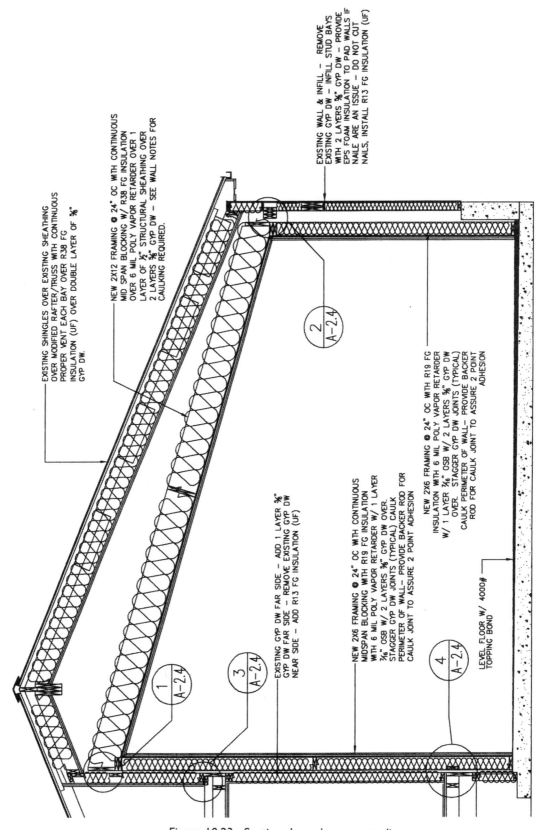

Figure 10.23 Section through garage studio.

A comment here regarding Figure 10.23; this is a drawing from a project I designed. The garage was oversized in this case, with rooms behind it and a trussed roof. We wanted to get some additional volume in the room, so we modified the trusses in order to achieve this. It is always possible to modify trusses or remove ceiling joists to create a cathedral ceiling if you really want to. But don't try this without first consulting with a structural engineer to make certain that the design is going to be adequate enough to provide all of the support required when you are finished with the work. It would be a real pain to invest all that money and then have the roof collapse, with the possibility of harming someone in the process.

In the case of this project, a ridge beam was added beneath the peak of the roof, with some temporary supports provided while the work was taking place. Then the existing trusses were modified (one at a time) with additional support placed beneath the upper chord of the truss. The support was attached to the truss with plywood gussets that were fastened using both construction adhesive and screws.

Hanger support was also added at the top and bottom of these members for bearing purposes. When this work was completed, the section of the roof forming the room was essentially carried by our new "rafters," which were capable of carrying the dead and live loads required by the building code for new construction. The remaining sections were then modified to re-create the truss bearing for that area of the building.

Masonry Construction

For those of you who live in areas where wood construction may be more expensive than masonry, or perhaps just prefer masonry construction to wood, or want to construct a hybrid of the two, masonry is indeed a viable option in the construction of recording studios.

When building Power Station New England, we used a combination of a masonry outer shell with wood-framed walls inside.

Masonry can be used very effectively to create isolating assemblies due to both its mass and the rigid frame it has when construction is completed. But you should understand that a single masonry wall will not get the job done for you. Whether the wall is a multi-wythe brick structure or constructed using hollow core cement or cinder block, once completed, it is a single leaf. The air spaces in this do not form a two-leaf system. Thus, two separate walls will be required for good acoustic isolation.

There is not a whole lot of detail I need to provide you here, but there are just a few points that I want to make. I say *no details are required* because this all boils down to solid walls and openings, and any good mason is very well versed in how to make both.

Use hollow masonry blocks for your construction—obviously, the larger, the better (12" blocks will create better isolation than 8" blocks due to greater mass and thickness).

Fill any cores required with structural grout and reinforcing rods to create structurally sound wall assemblies. Do not fill any cores with grout that do not require it for structural bracing; instead, fill these cores with dry sand.

It's important to make certain that any penetrations that are required for exterior lighting, alarms, red light systems, etc., are sleeved (PVC pipe is better for this than steel because it will never rust) with the sleeves grouted in place before the placement of sand in those cores.

Once the wall is finished, it is too late to add these because penetrating the cores at that point will release the sand, which, in turn, will destroy part of the wall's mass. A core that is grouted for structural purposes can always be cored after the grout is set to create a pathway for anything you need to pass through the wall, although it is always easier to install a sleeve first.

Sand versus Grout

The reason I recommend sand versus grout in those cores is this: although structural grout has plenty of mass (roughly 137pcf versus 92pcf for general-purpose construction sand), it has no damping effect.

If you take a masonry wall and tap on the grouted cores lightly with a hammer, you will clearly hear the tap on the other side of the wall. However, take that same wall and tap (with exactly the same force) on a sand-filled core, and the sound on the other side will be more muted when compared with the solid core.

This is due to the damping effect of the sand.

The little difference in mass is well made up for by the additional properties the sand brings into the picture.

As noted previously, you need two walls to make this work. If they are two masonry walls, the larger the air space, the greater the isolation will be. I would recommend a minimum of 8" between walls for this purpose.

If you really want to get serious with this type of assembly, I would recommend reading the following article. "Sound insulation of partitions in Broadcasting Studio Centres: field measurement data,"[3] which can be found here:

http://downloads.bbc.co.uk/rd/pubs/archive/pdffiles/architectural-acoustics/soundinsulationofpartitions.pdf

OK, you designed it, you detailed it, you built it, and you're happy with the isolation. Be proud of yourself for paying attention, but you aren't done yet.

This chapter is entitled "Putting It All Together" for a reason. It's going to take you right to the finish line.

Studio Treatments

You spent quite a bit of time looking at the choices you have for studio treatments in the last chapter. Now it's time to look at where they actually go.

Over the years, there have been several different concepts used for studio treatments. From early studios with near anechoic conditions to LEDE (Live End Dead End) studios (which became popular after the first LEDE room was constructed by Chips Davis in 1978[4]), everyone in the recording industry has searched for the ever-elusive perfect combination of room-to-treatment ratio. This is a search that continues today.

First a Bit of History

The 1960s: Dead Rooms—Very uncomfortable to work in and almost anechoic in nature.

The 1970s: Rettinger Rooms—The beginnings of the stereo generation, with people developing an interest in what was happening in the room. Good rooms had front walls made from stone and a lot of rear absorption. Raked sidewalls and ceilings helped control flutter echo. Rooms sounded very different from one another, and there were some serious problems with sounds within the spaces as well. An interesting side note here: M. Rettinger was an advocate of reverberant recording studios, utilizing rockwool and rigid glass boards, as well as slat treatments to maintain acoustics, going back into the 1950s[5].

The 1980s: Davis' LEDE Room with Reflection Control—Live End/Dead End really took off in popularity in the 1980s. These rooms were really the first real attempt to understand (and deal with) not only room sounds, but psychoacoustics as well. With the front of the room dead (almost anechoic) and the rear end of the room live, Davis, and those who followed him, found a new, fresh approach to studio monitoring environments. The use of reflection control helped to create an environment with better stereo imaging. Using diffusion on the back wall helped to break up room echo, while still maintaining room energy.

The 1990s: ESS (Early Sound Scattering)—A design for control rooms where reflections were uniformly random, such that they had no character to impose on the listening space. An ESS control room featured a highly diffusive front end that used diffusers to scatter early reflections. The remainder of the room was absorbent, and it generally used membrane panels for low-frequency control. These rooms were pretty live compared to older control rooms. They tended to have a flat frequency response and pretty good stereo imaging.

The 2000s: ? Where it goes from here is anyone's guess at this point in time.

Beyond History

In the anechoic control room, the goal was for none of the room to affect the sound of the recording, allowing the engineer to hear just exactly what came from the speakers alone. This method of room treatment created a very unnatural environment for the engineer (the human mind is used to hearing sound reflections), one in which the engineer tired rather quickly. This also made it very difficult to determine the amount of reverb to add to any track, as reverb was "eaten up" by room acoustics.

At the complete opposite of the spectrum, an untreated room creates problems with low-frequency modulation (as discussed in Chapter 2) and with higher frequency reflective sounds.

So you have two extremes—and somewhere between the two sits a perfect world. Or perhaps not.

Sound is to the ear as beauty is to the eye—thus, it is "of the beholder." If you did a survey of studio designers around the world and had them list the design they believed created the "best" studio, you would have almost as many different answers as you had designers. There really is no general consensus as to the best studio design (although all would agree that dead, almost anechoic is undesirable).

So one picks one's path and walks it.

My personal choice leans toward a Reflection Free Zone (RFZ) with plenty of bass trapping. This generally works well with small rooms, which usually do not have the room required for effective diffusion. So with that in mind, let's begin with some basics.

A good friend of mine (Ethan Winer) has a saying that I love: "A corner is a corner is a corner." Let's look at this as it relates to locations for bass trapping. He's absolutely right. You have already found (in Chapter 2) that bass builds up more strongly in corners and most strongly in trihedral corners. You also know by now that the biggest challenge facing you in small rooms is low-frequency modal (and non-modal) issues. So where do you start with the treatments you need?

In the corners of the room.

"But Rod," you say, "all of the corners are not available. For example, in the control room (we've been examining in this chapter), the corners by the door and the window into the iso-booth are useless for this purpose."

And in saying this, you're absolutely correct. So let's step outside of the box that locks most people into thinking "wall-to-wall intersections" when picturing "room corners" and look at this in three dimensions.

Figure 10.24 is a simple cube—could be a cardboard box, could be a room; either way, it has dihedral (two-point) and trihedral (three-point) corners. The numbers placed at various locations with the box indicate where two surface corners are available in the room, with the letters indicating available three-corner surfaces.

Figure 10.24
Available corners in a rectangular cube.

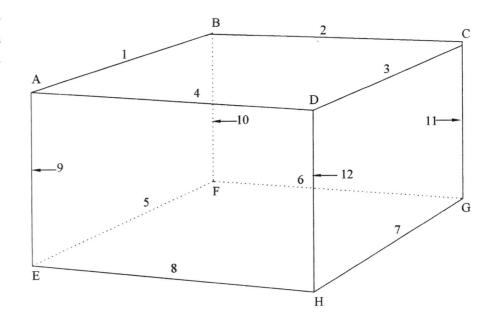

So in a simple cube, you have 12 dihedral corners, plus 8 trihedral corners, available for bass treatments. Now let's take a look at a simple 3D view of the control room you've been dealing with in this chapter. If you examine Figure 10.25, you'll see a total of 19 dihedral corners, plus 10 trihedral corners, available for treatment.

Figure 10.25
Control room corners.

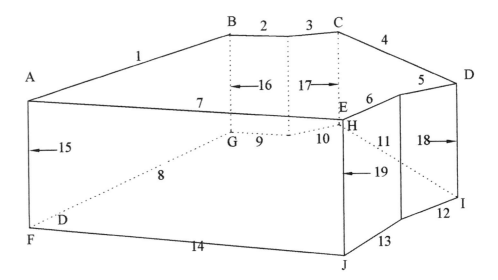

Note that six corners were not included in the list above; these are outside corners, which will not create a buildup of bass frequencies. Only inside corners contribute to this effect. So these corners were not included in either of the two or three face conditions.

Now, in the case of Figure 10.25, you know that there is a duct running below the ceiling, which runs from point A-E on line #7. There is also the beam that runs parallel and in front of line 2, crossing line 4.

However, if you look back at the reflected ceiling plan, you'll see some "balancing soffits" designed to maintain room symmetry, running from points A-B along line 1, and D-E along lines 5 and 6, which are perfect for bass traps.

You would also want to place a bass trap (floor to ceiling) in front of line 15 and add traps behind your speakers at each end of the window from lines 16.5 to 17 (Corner C) and 18 to 18.5 (Corner D). This gives you five locations within the room for bass traps to begin your treatments. With floor-to-ceiling and wall-to-wall traps (located at the soffit areas), that's a considerable amount of bass treatment to get you started.

The next recommendation would be to develop a cloud over the room to help handle ceiling reflections. The cloud does not have to cover every inch of the ceiling. It could be installed with a perimeter strip that ran 1' 6" from the edge of the soffits, creating a coffer effect for the ceiling (which also happens to work with the supply register locations). This is an effective room treatment, and it can also house some shallow, recessed, can-type lighting without the need to penetrate your ceiling. Although treating your ceiling helps you a little bit with low frequencies, it is really helping you much more with problems relating to higher frequency reflections.

You then have to examine your walls from the listening position to complete your RFZ, which means you have to determine your speaker and seating locations.

This is a small control room, and you'd probably be better off with nearfield monitors than anything else, but you'll work your way through that when you're all done building. Layout of listening position isn't all that difficult. You begin by being centered in the room left to right and with your ears sitting 38 percent of the room depth from the front wall.

Take a look at Figure 10.26 for a moment. It shows you the starting points.

Running a line down the center of the room, from top to bottom, is the first step. Then locate the listening position (38 percent back), which in this case is 7' 1 1/8" back from the front wall. Step 16" back from here to locate the mic for testing, and then lay out your speakers off this point, using an equilateral triangle. The speaker location ends up being centered on the 60° lines, 1' 9 5/8" from the front wall. The location of the speakers (in this case) was specifically determined to allow placement of treatments in the corners behind them. At this point, you can begin room testing or go directly to placement of room treatments.

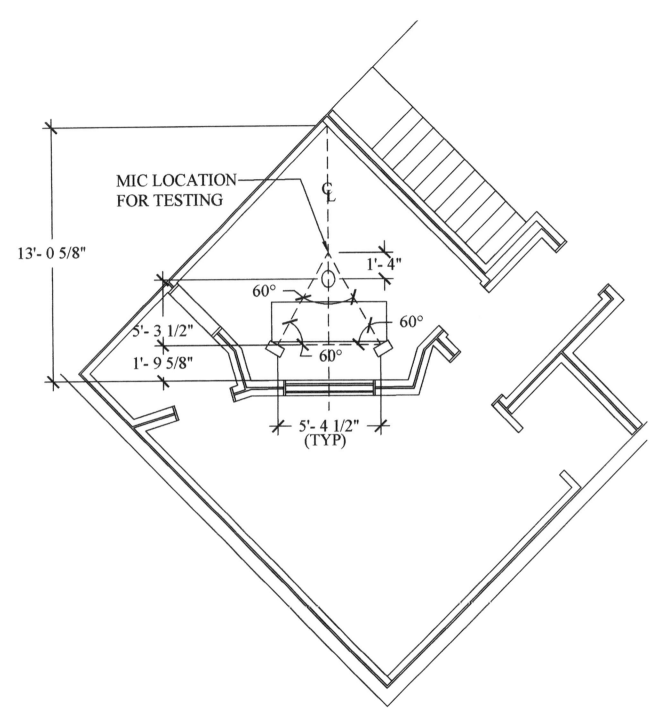

Figure 10.26 Seating and speaker location layout.

A personal thought on that—it is my belief (and that of a lot of my peers) that the treatments suggested to you in this book are the treatments required for small rooms (roughly 22 percent to 25 percent coverage of total room surface area). Although you can go through the exercise of room testing prior to treatments if you want, you will (99.99 percent of the time) end up in the same place at the end.

To save you from doing the math:

- ▶ Floor/ceiling areas are 118.5s.f. each
- ▶ Wall surfaces total 325.61s.f.
- ▶ Total surface area in room = 562.61s.f.
- ▶ Three-corner bass traps—6.83' high by 2' in width total 40.98s.f.
- ▶ Ceiling cloud totals 38.73s.f.
- ▶ RFZ totals 32s.f.
- ▶ Soffits total 38.42s.f. (this includes face and bottom of soffit)
- ▶ Total Treated Area = 150.13s.f.

These figures account for 26.68 percent of the total room surface.

Thus, I would not spend my time testing when I already knew the outcome. Having said that, how you proceed is entirely up to you. This is only said in relation to the room treatments I would originally plan for. It is a great idea before installing treatments to test the room for optimal speaker and listener locations before you begin room treatments, with measurements following that to determine if you need to "tweak" your room a little bit to get a great room versus a mediocre room.

It's now time to look at the same floor plan and consider actual trap locations. Let's take a minute to understand the need for an RFZ control room.

One issue with reflected sounds in a listening environment relates to something known as the Haas effect.

An acoustician by the name of Helmut Hass discovered a psychoacoustic effect, which indicated that a listener can hear two identical sounds, from two separate sources, and, if the subsequent arrivals are within roughly 15–25ms of the first, the mind will create the impression that two sounds came from the first source. If the second arrivals are later than this, then the ear hears two distinct sounds. The human mind will disregard the second sound if it arrives later than the outside edge of the Hass effect. This would be the difference between the mind hearing what it would describe as an echo versus reverb.

However, the European Broadcasting Union (EBU) recommends that levels of reflections earlier than 15ms relative to the direct sound should be treated to reduce them to at least 10dB below the direct sound for all frequencies in the range of 1kHz to 8kHz.

In terms of sound, 15ms of time represents 16.95 feet of travel distance. It's easy to see that in a room as small as this one, it would be virtually impossible to find any place in the room where a reflected sound could not find you within that travel distance. Just to be clear here—these speakers are sitting in this layout about 4' 3" away from your head, so the first 4' 3" of reflected travel doesn't affect the time delay—it begins after that.

How you determine where this will take place (which is where you will require treatments) is easy. You simply use a mirror. With you sitting in the listening position and the speakers in place, have a friend walk around the room sliding a mirror along the walls (and ceiling, if you decide not to use a cloud). Anywhere you can see a speaker in the mirror is a first reflection you have to deal with. This includes all walls of the room.

Figure 10.27 shows you where in this room that would be an issue.

In this layout, you can see that there are four locations where the walls are a potential issue. (Remember that you have already chosen the three corners for bass traps as shown, so any possible reflections from the front two corners are already being dealt with.) The total length of travel for line A is 12' 2", while the line B travel distance is 15' 3".

When you deduct the matching travel distance, your remainder is 7' 11" for A and 11' even for B. Both of these numbers are considerably below the 19.95' of travel that takes place in the 15ms interval and thus requires treatment.

Figure 10.28 shows you the final treatments in place.

One other thing, understand that creating a reflection-free zone is dealing with comb filtering, flutter echo, and other mid- and high-frequency room anomalies. It offers little toward dealing with modal activity all by itself. It takes the whole package to get it right.

While you are installing your treatments, make certain to test the room along the way. You want to find the right balance—sucking all of the bass out while leaving high frequencies lively is not what you're looking for, and neither is the reverse. You will consider the amount of testing you have to do to make this work well a real pain, but when all is said and done, it really will be a pain worth bearing. You get it right now, and you will have a lifetime enjoying it ahead of you.

Chapter 10 Putting It All Together

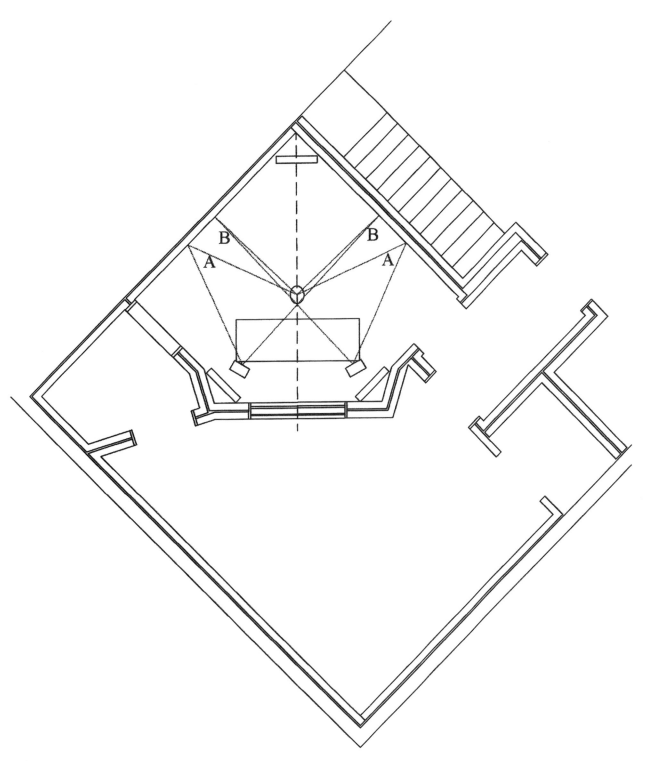

Figure 10.27 Creating an RFZ layout.

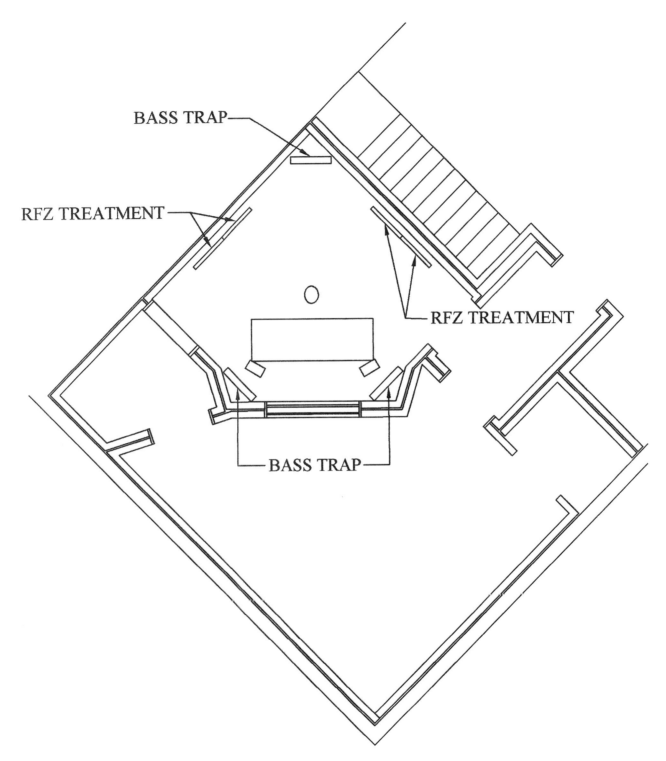

Figure 10.28 Room treatments.

The Finished Product

Well, there you go. You have a finished product and can now begin filling it up with gear. You worked long and hard to reach this point, and you deserve the enjoyment your room will bring you over the years.

Enjoy.

Endnotes

[1] The Span Calculator used was provided by Washington State University and can be found at http://timber.ce.wsu.edu/.

[2] From the National Design Specification for Wood Construction, National Forest Products Association, Washington, DC, 1997.

[3] Sound insulation of partitions in Broadcasting Studio Centres: field measurement data, BBC Engineering, Research Department, Engineering Division, October 1986.

[4] History and Development of the LEDE Control 1954 (B-5) Room Concept, by Chips Davis and Glenn E. Meeks, presented at the 72nd AES Convention, Anaheim, California, October 23–27, 1982.

[5] "Reverberation Chambers for Broadcasting and Recording Studios," Michael Rettinger, *Journal of the Audio Engineering Society*, January 1957, Volume 5, Number 1.

[6] Helmut Haas's doctorate dissertation presented to the University of Gottingen, Gottingen, Germany as "Über den Einfluss eines Einfachechos auf die Hörsamkeit von Sprache;" translated into English by Dr. Ing. K.P.R. Ehrenberg, Building Research Station, Watford, Herts., England Library Communication no. 363, December, 1949; reproduced in the United States as "The Influence of a Single Echo on the Audibility of Speech," *J. Audio Eng. Soc.*, Vol. 20, pp. 145–159, March 1972.

[7] "Listening conditions for the assessment of sound programme material: monophonic and two–channel stereophonic," EBU Tech. 3276, 2nd edition, May 1998.

CHAPTER 11

Myths and Legends

The Internet is a wonderful source of both information and misinformation. You'll find that a search for a particular item or phrase will return thousands, if not millions, of references. For example, a recent search for the term "acoustics," using a popular search engine, turned up about 16,500,000 references in a search only lasting 0.13 seconds.

That's sixteen million five hundred thousand...

Now that is amazing.

Just as amazing is that the vast majority of those references probably contain information that's faulty. Ferreting out what's real and what's false can prove to be troublesome to say the least. For example, the error in the calculation for Helmholtz slot resonators that you were reading about in Chapter 9, "Room Treatments," not only exists in some obscure Web sites, but also at some universities. How's that for making it tough?

My best suggestion to you, if you want to learn more about acoustics than what you've learned in this book, would be for you to buy some of the excellent books on the subject, such as *The Master Handbook of Acoustics*, by F. Alton Everest.

In addition, there are some excellent Web sites worth visiting, a few of which are hosted by some of the brighter minds in the world. The following are a list of some sites I visit on a regular basis:

- http://recording.org
- http://forum.studiotips.com
- http://musicplayer.com
- http://www.soundonsound.com
- http://johnlsayers.com
- http://pmiaudio.com

Note that the "studiotips" forum is strictly dedicated to acoustics, and the site from John L. Sayers is devoted to studio acoustics and design. The remaining forums have a fairly wide range of material available relating to many different aspects of the recording world, including recording techniques. Just beware of the "Internet experts," people whose only claim to fame and expertise rests on their ability to post something where everyone in the world can see it.

Some Popular Myths and Legends

What is a myth? The definition that affects us in some manner in acoustics follows:

Main Entry: **myth**
Pronunciation: 'mith
Function: *noun*
Etymology: Greek *mythos*

"A thing having only an imaginary or unverifiable existence."

Let's take a walk through some of the more popular myths relating to sound isolation and acoustic treatments making their way across the Internet today.

Fiberglass

Are fiberglass (and glass wool-fiber) products carcinogens? Or are they safe materials to use in home studios?

Well, it all depends on who tells the stories.

Background Information

In the early 1990s, it was announced that fiberglass had been classified as a "possible carcinogen," and that products used in the HVAC industry (such as fiberglass ductboard) could be putting people in danger. This was due to the possibility of small fibers breaking free (from the body of the duct) and entering rooms that were being treated by the HVAC systems. In those rooms, this material could then be taken into the body through the respiratory process and lodge in the lungs. The concern was similar to problems relating to asbestos.

Because I take the well-being of people very seriously, it was immediately announced that unprotected fiberglass ductboard would no longer be allowed on any project—residential or otherwise—that I was involved with. However, the installation of polymer-lined ductboards would be acceptable. This was due to the polymer lining's ability to hold glass fibers in place.

I followed the track of this investigation throughout the 1990s into the present and offer the following from the International Agency for Research on Cancer (IARC):

"Direct contact with fiberglass materials or exposure to airborne fiberglass dust may irritate the skin, eyes, nose, and throat. Fiberglass can cause itching due to mechanical irritation from the fibers. This is not an allergic reaction to the material. Breathing fibers may irritate the airways resulting in coughing and a scratchy throat. Some people are sensitive to the fibers, while others are not. Fiberglass insulation packages display cancer warning labels. These labels are required by the U.S. Occupational Safety and Health Administration (OSHA) based on determinations made by the International Agency for Research on Cancer (IARC) and the National Toxicology Program (NTP).

1994—NTP listed fiberglass as "reasonably anticipated to be a human carcinogen" based on animal data.

1998—The American Conference of Governmental Industrial Hygienists reviewed the available literature and concluded glass wool to be "carcinogenic in experimental animals at a relatively high dose, by route(s) of administration, at site(s), of histologic type(s) or by mechanism(s) that are not considered relevant to worker exposures."

1999—OSHA and the manufacturers voluntarily agreed on ways to control workplace exposures to avoid irritation. As a result, OSHA has stated that it does not intend to regulate exposure to fiberglass insulation. The voluntary agreement, known as the Health & Safety Partnership Program, includes a recommended exposure level of 1.0 fiber per cubic centimeter (f/cc) based on an eight-hour workday and provides comprehensive work practices.

2000—The National Academy of Sciences (NAS) reported that epidemiological studies of glass fiber manufacturing workers indicate "glass fibers do not appear to increase the risk of respiratory system cancer." The NAS supported the exposure limit of 1.0 f/cc that has been the industry recommendation since the early 1990s.

2001—The IARC working group revised their previous classification of glass wool being a possible carcinogen. It is currently considered not classifiable as a human carcinogen. Studies done in the past 15 years since the previous report was released do not provide enough evidence to link this material to any cancer risk."[1]

I also want to reference the following studies, which draw the same conclusions:

Agency for Toxic Substances and Disease Registry. Technical Briefing Paper: "Health Effects from Exposure to Fibrous Glass, Rock Wool or Slag Wool." Agency for Toxic Substances and Disease Registry, U.S. Department of Health and Human Services, Atlanta, GA, 2002.

National Toxicology Program. *Glasswool in Report on Carcinogens*, 9th Edition. U.S. Department of Health and Human Services, National Toxicology Program, Research Triangle Park, NC, 2001.

"Man-Made Vitreous Fibres, Special-purpose glass fibres such as E-glass and '475' glass fibres (Group 2B), Refractory ceramic fibres (Group 2B), Insulation glass wool (Group 3), Continuous glass filament (Group 3), Rock (stone) wool

(Group 3), Slag wool (Group 3)," VOL.: 81 (2002), International Agency for Research on Cancer, World Health Organization, Lyon, France

The above information notwithstanding, there are still those out there, including Web sites, which would appear to be (otherwise) "legitimate" sources of information, that make serious claims to the contrary. Simply put—they are wrong.

My Advice

So, for you, here is my advice when it comes to fiberglass.

When working with any glass fiber products, wear long-sleeved shirts with the collar and sleeves buttoned. Wear a dust mask and safety goggles.

Under normal conditions, once the glass panels are fabric wrapped, the fibers should be contained within.

When work is completed, use a vacuum cleaner to clean up any stray particles. Do not use a broom to clean up fiberglass dust, as this will raise and redistribute a large amount of particles. If available, a vacuum cleaner with a HEPA filter is the best way to go.

Common sense dictates that you wouldn't want to dump this stuff in your body, but with a little bit of care, it's perfectly safe to work with and have in a home environment.

Egg Crates and Other Great Acoustic Treatments

It's amazing the number of otherwise intelligent people who make comments such as:

"If you want great acoustic treatments for cheap money, use egg crates."

"If you want great acoustic treatments for cheap money, just stick used carpet on your walls."

"If you want great acoustic treatments for cheap money, there's nothing like packing foam, after all—foam is foam."

"If you want great acoustic treatments for cheap money, put some mattresses up against your walls."

(You could fill up a book with comments like these.)

Always followed by:

"I know, because I did it, and now my room is perfect."

Let's take a minute and examine these claims that they say work.

Although I have never found any test data for mattresses or packing foam, there is test data for egg crates and carpet. Acoustics First Corporation (formerly Alpha Audio) had egg crates tested at Riverbanks back in 1988, and here are the results:

RIVERBANK ACOUSTICAL LABORATORIES
OF IIT RESEARCH INSTITUTE
REPORT

1512 BATAVIA AVENUE
GENEVA, ILLINOIS 60134

312/232-0104
FOUNDED 1918 BY
WALLACE CLEMENT SABINE

FOR: Alpha Audio

ON: Egg Crates

CONDUCTED: 28 March 1988

Sound Absorption Test
RAL™-A88-80

Page 1 of 4

TEST METHOD

The test method conformed explicitly with the requirements of the ASTM Standard Test Method for Sound Absorption and Sound Absorption Coefficients by the Reverberation Room Method: ASTM C423-84a and E795-83. Riverbank Acoustical Laboratories has been accredited by the U.S. Department of Commerce, National Bureau of Standards under the National Voluntary Laboratory Accreditation Program (NVLAP) for this test procedure. A description of the measuring technique is available separately. The microphone used was a Bruel & Kjaer serial number 1330658.

DESCRIPTION OF THE SPECIMEN

The test specimen was designated by the client as egg crates. The overall dimensions of the specimen as measured were 2.34 m (92 in.) wide by 2.63 m (103.5 in.) long and nominally 5.1 cm (2 in.) high. The specimen was tested in the laboratory's 292 m^3 (10,311 ft^3) test chamber. The weight of the specimen as measured was 4.5 kg (10.0 lbs) an average of 0.73 kg/m^2 (0.15 lbs/ft^2). The area used in the calculations was 6.14 m^2 (66.1 ft^2). The room temperature at the time of the test was 21°C (69°F) and 61% relative humidity.

PRECONDITIONING

The specimen was held at least 24 hours under the test conditions of 21°C (70°F) and 60% relative humidity.

MOUNTING A

The test specimen was laid directly against the test surface.

THE RESULTS REPORTED ABOVE APPLY ONLY TO THE SPECIFIC SAMPLE SUBMITTED FOR MEASUREMENT. NO RESPONSIBILITY IS ASSUMED FOR PERFORMANCE OF ANY OTHER SPECIMEN.

NVLAP ACCREDITED BY DEPARTMENT OF COMMERCE, NATIONAL VOLUNTARY LABORATORY ACCREDITATION PROGRAM FOR SELECTED TEST METHODS FOR ACOUSTICS.

RIVERBANK ACOUSTICAL LABORATORIES
OF IIT RESEARCH INSTITUTE
REPORT

1512 BATAVIA AVENUE
GENEVA, ILLINOIS 60134

312/232-0104
FOUNDED 1918 BY
WALLACE CLEMENT SABINE

Alpha Audio

RAL™-A88-80

28 March 1988

Page 2 of 4

TEST RESULTS

1/3 Octave Center Frequency (Hz)	Absorption Coefficient	Total Absorption In Sabins	% Of Uncertainty With 95% Confidence Limit
100	0.00	0.02	0.95
** 125	0.01	0.95	0.93
160	0.00	0.00	0.97
200	0.07	4.62	0.96
** 250	0.07	4.90	1.10
315	0.07	4.44	1.08
400	0.13	8.60	1.08
** 500	0.44	29.15	0.78
630	0.73	48.30	0.71
800	0.74	48.64	0.69
** 1000	0.61	40.59	0.70
1250	0.52	34.51	0.68
1600	0.46	30.27	0.74
** 2000	0.48	31.88	0.71
2500	0.58	38.17	0.72
3150	0.59	38.73	0.72
** 4000	0.69	45.38	0.65
5000	0.82	54.14	0.66

NRC = .40

THE RESULTS REPORTED ABOVE APPLY ONLY TO THE SPECIFIC SAMPLE SUBMITTED FOR MEASUREMENT. NO RESPONSIBILITY IS ASSUMED FOR PERFORMANCE OF ANY OTHER SPECIMEN.
NVLAP ACCREDITED BY DEPARTMENT OF COMMERCE, NATIONAL VOLUNTARY LABORATORY ACCREDITATION PROGRAM FOR SELECTED TEST METHODS FOR ACOUSTICS.

RIVERBANK ACOUSTICAL LABORATORIES
OF
IIT RESEARCH INSTITUTE

1512 BATAVIA AVENUE
GENEVA, ILLINOIS 60134

REPORT

312/232-0104
FOUNDED 1918 BY
WALLACE CLEMENT SABINE

Alpha Audio

RAL™-A88-80

28 March 1988

Page 3 of 4

TEST RESULTS (con't)

The percentage of uncertainty for the required 95% confidence limits indicated above must fall within the prescribed limits designated in par. 13.2 of ASTM C423-84a. It states that for the absorption of the reverberation room containing the specimen the testing laboratory shall obtain data with less than 4% uncertainty at 125 (hertz) and 2% uncertainty at 250, 500, 1000, 2000, and 4000 (hertz). The method of calculation is described in ASTM STP 15D and outlined in section 13 of the standard.

The noise reduction coefficient (NRC) is the average of the coefficients at 250, 500, 1000, and 2000 Hz, expressed to the nearest integral multiple of 0.05.

Reviewed by _Diane C. Perrone_   Submitted by _Peter E. Shaw_
Diane C. Perrone   Peter E. Straus
Senior Technician   Senior Technician

THE RESULTS REPORTED ABOVE APPLY ONLY TO THE SPECIFIC SAMPLE SUBMITTED FOR MEASUREMENT. NO RESPONSIBILITY IS ASSUMED FOR PERFORMANCE OF ANY OTHER SPECIMEN.

NVLAP ACCREDITED BY DEPARTMENT OF COMMERCE, NATIONAL VOLUNTARY LABORATORY ACCREDITATION PROGRAM FOR SELECTED TEST METHODS FOR ACOUSTICS.

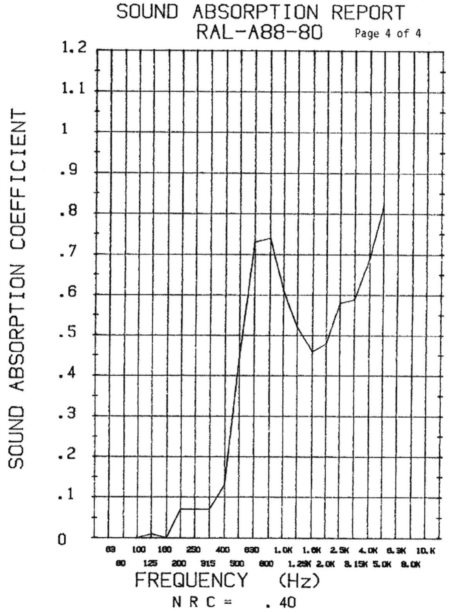

As you can see from the above test report, in the lower frequency ranges in particular, egg crates offer little value for absorption. At best, they can knock down a bit of mid/high frequencies, but this will leave your room muddy in the end.

Figure 11.1 is a side-by-side comparison of the absorption coefficients of egg crates and Auralex 2" Sonomatt Eggcrate-style foam.

	1/3 Octave Center Center Frequency (Hz)	Egg Crates	Sonomatt	Variance	Variance as Percentage
	100	0.00	0.08	0.08	0.00%
**	125	0.01	0.13	0.12	7.69%
	160	0.00	0.14	0.14	0.00%
	200	0.07	0.2	0.13	35.00%
**	250	0.07	0.27	0.20	25.93%
	315	0.07	0.35	0.28	20.00%
	400	0.13	0.47	0.34	27.66%
**	500	0.44	0.62	0.18	70.97%
	630	0.73	0.75	0.02	97.33%
	800	0.74	0.85	0.11	87.06%
**	1000	0.61	0.92	0.31	66.30%
	1250	0.52	0.96	0.44	54.17%
	1600	0.46	1.01	0.55	45.54%
**	2000	0.48	1.02	0.54	47.06%
	2500	0.58	1	0.42	58.00%
	3150	0.59	1.02	0.43	57.84%
**	4000	0.69	1.02	0.33	67.65%
	5000	0.82	1.06	0.24	77.36%
	Mounting	A	A		
**	Values reported as octave-band absorption coefficients in accordance with ASTM C423.				

Figure 11.1 Absorption coefficient comparison between egg crates and Auralex Sonomatt.

Note that from 200Hz and down, egg crates offer little or no value for sound absorption. There was some small value at 250, 315, and 400Hz, although you would use three to five times as many egg crates to achieve the same value as the Auralex Product. Again, almost no value from 500 through 800Hz. At 630Hz, they are almost identical, but then the numbers begin to fall off again—with the Auralex product performing nearly two to four times more efficiently than standard egg crates.

Even more telling is the percentage of confidence in the test results themselves. The last page of the report gives the standard ASTM used at the time of the testing of the egg crates. The requirement for uncertainty at 125Hz was 4% and was only 2% for 250, 500, 1,000, 2,000, and 4,000 (hertz). The intent of this was to demonstrate that there was a 95% certainty that the test results could be duplicated at this or any other test facility. Yet, the only frequencies that exhibit certainty are the low frequencies, i.e., 125 and 250Hz. The remaining frequencies fail the repeatability requirements for uncertainty. And you have basically no absorption value at 125 and 250Hz.

So the myth is put to bed. Even if the uncertainty factor did not come into play, the more material you place on the wall, the more you suck the mids and highs out of the room, leaving the bass modal issues, which you know are the biggest issues in small rooms.

Even if you got all of the egg crates for free, after the cost of treating them so they would be flame retardant, if you compare the total costs per sabins for performance to achieve the results you require, you'll find that this is not an effective means of room treatment.

Oh, one other thing regarding egg crates—the people who tell you how great they are will say that part of the reason they actually work better than test results indicate (because they can "hear the difference," mind you) is due to the fact that they are diffusors as well as absorbers.

Well, you've studied the concepts behind diffusors and know that these perfectly symmetrical little egg carriers wouldn't work well for that purpose (they are very narrow in the band they would diffuse based on their geometry), so you can straighten them out about that if they try to sell you on the subject.

Carpets

Carpet, when placed on the walls of studios, can attenuate mid and some higher frequencies (the same as it will do on a floor), but is (again) not the answer for low frequencies.

This, too, is coupled with the fact that carpet is not made to be put on walls (with the exception of some Berber carpets that have test data certifying them for that purpose).

Figure 11.2 is from the Carpet and Rug Institute regarding the acoustic value of carpet:

Note that the NRC (noise reduction coefficient) for carpet is fairly low, especially when you consider that the only carpet you can put on walls is Berber carpet (or at least those Berber carpets that have been tested and approved). Berber carpet comes under the variable heading "increasing pile weight/height relationships in woven wool loop pile carpet." Note that the NRC for this carpet is only 30 to 40 or roughly half that of the Auralex Sonomatt, and you still need to think about ease of installation.

Test Series A-1 Carpet was placed directly on the concrete floor of the test chamber.

Test Variables	Pile Weight oz/sy	Pile Height Inches	Surface	NRC
Identical construction, different manufacturer	44	.25	Loop	.30
	44	.25	Loop	.30
	44	.25	Loop	.30
Identical construction, different pile surfaces	35	.175	Loop	.30
	35	.175	Cut	.35
Pile weight/height relationships in cut pile carpet	32	.562	Cut nylon	.50
	36	.43	Cut acrylic	.50
	43	.50	Cut wool	.55
Increasing pile weight/height relationships in woven wool loop pile carpet	44	.25	Loop	.30
	66	.375	Loop	.40
	88	.5	Loop	.40
Increasing pile weight (pile height constant) in tufted loop pile carpet	15	.25	Lop nylon	.25
	40	.25	Loop wool	.35
	60	.25	Loop wool	.30
Varying pile height (pile weight constant) loop pile with regular back		.125	Loop	.15
		.187	Loop	.20
		.250	Loop	.25
		.437	Loop	.35
Varying pile height (pile weight constant) loop pile with foam back		.187	Loop	.25
		.250	Loop	.30
		.312	Loop	.35
		.437	Loop	.40

Observations: A-1
1. Carpet tested in this program, which were laid directly on concrete, had NRCs ranging between .15 and .55.
2. It was found that when manufacturers met identical specifications, their fabrics have the same NRCs. However, the sound absorption coefficients at individual frequencies varied somewhat.
3. Cut pile carpet, because it provides more "fuzz", provides a greater NRC than loop pile construction in otherwise identical specifications.
4. As pile weight and/or pile height increases in cut pile construction, the NRC may not change substantially.
5. Increasing pile weight while increasing or holding pile height constant in loop pile construction resulted in sound absorption "topping out" because the surface does not change in absorptivity at higher frequencies.
6. Increasing pile height while holding pile weight constant in loop pile fabrics results in improvements in absorption. Loop pile carpoets average NRC values of .20 to .35.
7. Foam backed loop construction resulted in an increased NRC value compared to conventional secondary backed carpet.

CRI Technical Bulletin: Acoustical Characteristics of Carpet (0/00)

Figure 11.2 Noise-reduction coefficient on carpet.

Understanding NRC is easy—you take the average sound absorption coefficient measured at four frequencies: 250, 500, 1,000, and 2,000Hz expressed to the nearest integral multiple of 0.05. So it's easy to see that you aren't getting any great ratings at those frequencies if your NRC is only 30 to 40.

Next, you need to read this (also from the Carpet and Rug Institute), regarding the placement of carpet on walls:

"Carpet is manufactured for use as a floor covering, and installation on other surfaces, such as walls, is not recommended. Many carpet manufacturers will not assume any liability, real or implied, when carpet is applied on surfaces other than floors."

Interestingly enough, most insurance companies won't assume liability either. If you place carpet on walls, and it is determined that it contributed in any way to a fire, that is a perfect opening for your insurance company to walk away from you.

Once again—no low-frequency benefit—and little benefit for what it does provide, especially when you consider the risk at which you are placing yourself.

Packing Foams

We've touched on this earlier in the book. These foams are only to be used for what they are made—shipping packaging. They do not attenuate sound the same way that products made specifically for that purpose do. I don't care what your ears told you—your ears are wrong.

If, after all that's happened with nightclubs and packing foams in recent years, anyone still wants to use these in their room(s), they must be crazy. Once again, your insurance company will drop you like a hot potato, and you can forget about collecting for a fire if that product is installed and contributed to a fire in any way.

Go back to Chapter 9 if you have any questions remaining and re-read the information there relating to foam products.

Mattresses

Yup, some people absolutely swear by these for use in studios as bass traps. They also make claims that they will help you big-time with sound isolation.

First, as you well know by now (having read the book to this point), isolation requires mass, which is sorely lacking from a mattress. Low-frequency attenuation requires mass and air, only one part (of which) is provided by a mattress. Mattresses will treat mid- and high-frequency transmissions (seem to be seeing a trend here with that), but, once again, the biggest problem in your rooms is going to be low-frequency modal and non-modal activities.

Second, safety is an issue again. Although these won't take to flame quite the same way as packing foam, they will smolder for a long time before a fire actually takes off. You could well have a fire start in a mattress and not know it for hours after it began.

As always, because the acoustic benefits you might get from the material aren't really helping you where you need it (as well as weighing in the safety factor), don't even bother heading in this direction.

Please Buy Our "Soundproofing" Materials

Beware of companies selling "soundproofing" foams, fiberglass, insulation, etc. There are a fairly large number of companies out there that advertise what are actually sound attenuation products as "soundproofing." Because they don't have a clue what their product really is, or how to market it properly, I have serious concerns about whether they produced a product worth purchasing. Let the buyer beware.

The Myth of Soundproofing

A note on the general concept of soundproofing itself, as well as companies that tell you they can soundproof your rooms, homes, etc.

Soundproofing doesn't exist.

Possibly the quietest recording facility in the world has to be Galaxy Studios located in Belgium. The design is wild and crazy (credit to Eric Desart and Gerrit Vemeir) when you consider that 100.7dB in isolation is consistent throughout the entire facility. It is said that co-owners Wilfried and Guy Van Baelen are happy to step into one of the studios, close the airtight, concrete-filled door behind them, pull out a starter's pistol (whose report comes in around 125dB), and fire it. And if you hadn't seen someone pull the trigger, you'd never know it had happened. Understand that this level of isolation had never been accomplished before in any commercial facility.

Interesting thing is—that isn't soundproof, it is just darn quiet, possibly quiet beyond belief, but not soundproof.

For example, a .375 cal. centerfire rifle, with an 18" barrel and a muzzle brake will produce 170dB upon firing.[3] That would be heard in an adjacent room. (By the way, that would be one uncomfortable rifle to fire without hearing protection since the threshold of pain is 130dB.)

This doesn't mean that the studio isn't quiet—it just means it isn't soundproof.

Avoid any company trying to sell you "soundproofing." The fact that they even use the term is evidence of their lack of knowledge on the subject.

Close Is Good Enough

This is the area where people get caught. And it is a costly mindset for someone trying to achieve isolation.

The details outlined in this book are there for a reason; they ensure that you make it from point "A" to point "B." They are not there to cost you additional monies, but rather to ensure that failures due to the "human factor," and possible deficiencies in materials, do not come back to haunt you in the end.

Thus, in a perfect world, a single layer of acoustic caulk would ensure that there would be no sound leakage through the perimeter of a wall assembly, and yet in here I recommend a caulk joint at each layer of drywall on the edges. The intent of this redundancy is a backup to cover the human factor or a product deficiency.

Let's examine both of those for a moment.

The Human Factor

Human performance has certain capabilities and limitations that affect the outcome of work produced. When human beings make an error, they will assess the work and make a determination that the error is either acceptable or not acceptable. For example, a piece of drywall cut partly out of a square would be typically viewed as being acceptable, whereas a piece of wood trim (intended for a clear finish) not fitting properly would generally be viewed as being not acceptable. These decisions are easy to make, because in one case the drywall joint will be taped over and thus covered up, while the wood trim will be exposed for all to see.

Therein lies part of the problem.

People tend to view things that aren't seen in the finished product as being not quite as important as the finished product itself. I've witnessed this time and time again during my many years in the construction industry. For example, there are framing carpenters who believe that framing doesn't have to be cut perfectly square and tight-fitting because it gets covered up and you never see it. There are trim carpenters who don't worry about painted trim fitting like it was cut with a razor blade, because the painters can just caulk the joint when they finish it. No one will know or see it in the finished product.

After all, close is good enough.

The fact is, especially in the case of sound-isolating construction, nothing could be further from the truth. It's this mentality that is the cause for walls that were designed with a particular isolation in mind to actually have field ratings of 20dB less or worse. This happens in more cases than one might believe.

But what the heck, close is good enough.

How does one take the human equation out of the picture? Through the use of redundancy. You think "redundancy" because the odds of a person getting sloppy, making a mistake in *exactly* the same location, in the same *exact* manner, in an additional application, are pretty slim. They may very well get sloppy—they may very well make a mistake—but it will not be in the exact same location.

So, having workers stagger the joints in each and every layer of drywall protects you from bad seams. Making them caulk each and every edge of those layers moves any areas of imperfection around the perimeter, with good joint areas protecting weak ones. While we're on the subject of good caulk joints versus weak caulk joints, let's look at the issue of product deficiency. I wish I had a dime for every time I've seen a caulk joint fail. Sometimes, they failed based on sloppy workmanship, and sometimes they failed when the workmanship was perfect.

Product Deficiency

Every product manufactured goes through a QC (Quality Control) process, and all of those QC processes have tolerance ranges, which are the maximum/minimum size, amount of chemical additive, etc. that can be used in the process prior to the product developing a failure. For an engine, it might be the max/min piston diameter vs. the max/min cylinder diameter. So, if the cylinder diameter is below the minimum allowable, then the max diameter piston won't fit within the chamber. By the same token, if the cylinder is over the maximum limit, then the smallest allowable piston might not create enough compression for the engine to operate properly.

The same goes with the chemical compounds that create the bonding capacities with caulk. There are very small variances allowed in the mixing process that will maintain the bonding and curing capacity of the products. But, if the ratios of the mixture are outside of these, the caulk will fail.

So what happens when we take the human factor into consideration with the QC process in these manufacturing plants? Someone in QC decides that an additional $1/1000$ of an inch (beyond allowable tolerance) isn't enough of a variance to throw out a run of 10,000 pistons, which end up being installed in some engines where the cylinders were at the maximum allowable tolerance—and you just bought yourself a lemon that never runs right.

In the case of caulk, someone in the QC process decides than an additional $1/1000$ of an ounce of a particular chemical (beyond design tolerance) isn't enough of a variance to justify throwing away 10,000 tubes of caulk—and you just bought some tubes that are going to fail.

But what can you expect, after all—close is good enough.

What makes this interesting is that it isn't generally 100% of the run that's out of tolerance. If the tolerance at the beginning of the run were right at the edge when that particular lot was being produced, then maybe only 5 to 8% are out of tolerance. If that were the case, would you throw away stock knowing that anywhere from 92 to 95% of the stock was good? Well, some people in QC would and some wouldn't. (I know this for a fact because my father used to work in QC and forever had people angry with him because he would reject an entire lehr of glass bottles because they slipped over the allowable tolerances. People would say to him, "If 'so and so' had inspected this, he would have let it go out, " to which Dad would reply, "Then maybe you should have had him inspect it, but it isn't going out with my stamp on it."

So you pick up 100 tubes of caulk at your local lumberyard, 5 cases from that same lot, and anywhere from 5 to 8 tubes are going to have joint failures. Let me clarify this by saying that I am not telling you that for every case of caulk you buy you are going to have five to eight tubes of bad caulk. This is just an example of what could happen with a bad lot that slipped through the QC process.

How does one take product failure out of the equation? Once again (in the case of caulk)—redundancy. If you caulk only one layer of drywall, then any failure of the caulk joint affects you directly. But calking each and every edge of those layers moves any areas of imperfection around the perimeter, with good joint areas protecting weaker ones.

Trust me, invest the money and go the extra mile. It is that commitment to excellence that is going to guarantee success rather than failure.

Trust Me—I Really Have Heard It All

I don't know why, but I still find myself amazed every time I hear people spouting the same incorrect data, it seems like someone, somewhere, would finally get it right, but the misinformation on the Internet just seems to continue to make its rounds—almost as if it had a life all its own.

People claim that you can get great (cheap) isolation by putting carpet down, plywood over that, then a layer of gypsum drywall, another layer of plywood, and then finishing off with some pad and carpet (a great recipe for creating a feeding ground for mold, mites, and insect infestation). Others who say that loosely hanging a sheet of drywall at a slight angle in front of an existing wall will create an isolating panel that will increase the TL value of your wall. They tell you to make sure to put some fluffy insulation in between the existing drywall and the new floating sheet (no studs because it just rests against the base of the wall and a wood molding at the top) so that it not only acts as an isolator but also as a bass trap. Of course, they can't tell you how it will perform, just that they did it and they "know it's fantastic."

And with the misinformation come the "acoustic experts" that start up companies. They are the people who sell isolation products that (when installed in accordance with their directions) take walls that were constructed in accordance with the building codes and turn them into walls that no longer conform to the code.

One company I found has products that require you to remove the drywall and then create their version of a staggered wall. (I will not mention a name here, just the concept, if you run across anyone selling products like this, run like the wind). You do this by building out every other stud so that your new drywall is bridging the studs in between without touching them.

Not a bad concept, but it does pose a problem. The Building Code (at least here in the States, you should verify the requirements in your own country if you don't live here) has requirements for the fastening of drywall for both bearing and nonbearing walls. The Code reference is to a document entitled GA-216, which is a publication from the Gypsum Association that outlines the standards for the installation of drywall. It covers everything from one-layer walls to multilayered walls, from mechanically fastened to laminated panels.

One of the things it does is to specify the maximum stud spacing for walls receiving drywall panels. These requirements are dependent on the wall type (bearing/nonbearing), whether the wall is wood or metal studs, and what the thickness of the drywall is.

There are situations where the maximum stud spacing is 12". Other times it is 16", and almost as often (as 16") it is 24". But it is *never* more than 24".

Because the vast majority of studs are spaced either 24" or 16" on center, that means that under the best of circumstances (with a wall framed 16" on center) your connections would be on 32" centers to use this product as designed by the manufacturer.

This is a code violation, but they never mention this on their Web site.

In addition to that, I spoke with a person who was almost a client of theirs (they claim almost magical improvements when their product is used), and they told him this would help him to stop the noise coming from his next-door neighbor's apartment. Which means they were not only telling him to take a perfectly legal wall assembly apart to create the code violation I noted above, but also to create another code violation by modifying a rated, tested fire-separation assembly that is required under the code.

In some states doing this would actually be a criminal act. (Connecticut is one of those states.) And in most states doing this could lead to criminal charges if the act created a situation where a fire broke through the wall and someone was injured (or died) in part or whole due to the act of destroying the fire barrier.

Another problem that could arise due to this is the possibility that the wall in question may well be a shear wall required as a part of the engineering to help hold the building up when faced with strong winds or seismic events.

This may or may not be an issue in single family homes (depending on the size of the home and it's location), but it is very commonplace for this technique to be used in apartment buildings, and quite often the walls in question are the party walls separating apartments from each other, or from hallways.

The company in question is obviously a start-up deal from a group of people who thought they had a good idea and figured it had to work. Heck, they might even have paid to have it tested acoustically. (The acoustical lab doing the testing doesn't worry about code issues. You ask them to test something, they test it, you pay them, and you get the report, but none of that makes it code compliant, or safe.)

If it seems too good to be true, it probably is. Make it a point to check with a building official, an architect, or an engineer before you start tearing something apart because some company told you that their product would work.

These are just a couple of examples of things to walk away from. Just remember that for every real company out there, there are a dozen that do not have a clue.

The Final Word on Myths

The bottom line here is: "Beware the Acoustical Myth."

There are a lot more fallacies and misconceptions in acoustics than could be presented in this chapter. In fact, an entire book could be devoted to the subject, but hopefully, you get the idea.

The examples given in this book are intended to help you avoid some specific problems (or concerns) with your studio design and construction. They are an attempt to illustrate the dangers in believing everything you see or read. Whether you read it in a magazine or on some Web site on the Internet, stick with what's tried, true, and tested. You'll never be sorry you did.

Anytime an acoustical myth can be identified and replaced with a little common sense or objective proof, acoustics as a science becomes less mysterious, and one less acoustical "truth" will be preached as gospel.

Endnotes

[1] IARC Monographs Programme Re-evaluates Carcinogenic Risks From Airborne Man-Made Vitreous Fibres, Press Release No. 137, 24 October 2001, International Agency for Research on Cancer, World Health Organization, Lyon, France.

[2] Riverbank Acoustical Laboratories, Sound Absorption Test RAL®–A88-80, 28 March 1988, reprinted with permission of Acoustics First Corporation.

[3] Documented by Dr. Krammer, Ph.D., Ball State University, Muncie, Indiana.

CHAPTER 12

Codes, Permits, and Special Needs

I feel the need to spend at least some time with you discussing building codes, permits, inspections, and engineering.

Building Officials

A large number of people I come in contact with who are doing home studio construction tend to view building inspectors as evil ogres bent on stopping them from doing what needs to be done to make their studios right. Nothing could be farther from the truth.

Building officials are simply doing their job, which is to ensure that a minimum standard of construction, which has been established to help provide safe buildings for people to occupy, is met.

When you walk into an official's office with something totally outside of their experiences, they (rightfully) look very hard at what you propose to do. They probably will require some documentation to reassure them that this meets (or exceeds) those standards established in the building codes. They will also perform periodic inspections on the work to ensure that the actual construction meets those same standards.

I have dealt with this for over 30 years, and you would be amazed at the number of times that an inspector notices a code violation on work done by professionals in their field. It could be something as simple as too many wires in an outlet box (which could create heat buildup within that box and cause a fire) to an incorrect nailing pattern used on a shear wall (which could cause a collapse of the building under wind or seismic loads).

I cannot stress enough the importance of these people and the jobs they do, and I recommend that you always pull a permit for any work that you intend to perform.

Structural Analysis of the Proposed Work

I have also found myself alarmed by the number of people who have posted their designs for construction on floors above grade that have not bothered to contact structural engineers to verify that the existing structure could take the additional loads they would be imposing on that structure. By "floors above grade," I'm referring to first floors with basements below, second floors of buildings, and so on.

On more than a few occasions, after posting my concerns, I have had the parties contact me and thank me for stressing the need for them to hire an engineer to review their design intent. The engineer assured them that they could *not* do what they wanted and that the structure could *not* carry the additional load.

My insistence that they do this not only saved them a lot of money, but it also may have saved a life in the process. One thing about structural collapses is that generally they do not give any warning. One moment an overloaded joist is seemingly fine, and the next moment it's gone.

Please do not add structural loads to buildings or cut and remove structural members without first checking with a structural engineer to verify that you can do this safely. Another thing that you need to understand is that my methods, although tried and true, do not and cannot cover each and every condition that exists in the world. Thus, you really need to consult with professionals in your area (I include building officials in that category) to make certain that you aren't going to damage your structure (at worst), or throw your money away while in the process of constructing your studio.

Let's take a look at a couple of examples to show you how important this really is. Understand that these are not the only times this matters, but rather are indicative of the need in general. One could fill a book with examples and still never scratch the surface of *all* of the problems that could arise.

Problems Relating to the Bearing Capacity of Earth

Earth has certain bearing capacities, which are different from area to area throughout the world, and can't be covered in this book. There are areas of the world that construct buildings above materials, which are referred to as *expansive soils*. These expansive soils are considered one type of unsuitable material for load-bearing construction to build on. There are other materials considered unsuitable—for example, earth containing large amounts of vegetation, structures built over old dump sites, and earth containing large amounts of loamy materials, just to mention a few.

For the purpose of this discussion though, let's look at expansive soils. Expansive soils are earth generally containing large amounts of clay by volume, which can expand and contract greatly, depending on the levels of the water table in that locale. Often, the elevation of the water table varies by season.

When buildings are built in areas with expansive soils that contain varying water levels, then the structure is generally supported through the use of either piles or caissons. Piles are supports made of either wood, concrete, or steel, which are driven into the ground with a machine called a *pile driver*. These supports are often driven to bedrock, although they may be designed to support the load strictly through the friction that the earth places on the side surfaces of the pile.

There are then concrete pile caps poured above these piles with a series of concrete grade beams that connect the piles to form the building's foundation. In this manner, the piles carry the entire building load without any transfer of load to the unsuitable soils that surround the building. Figure 12.1 depicts a simple pile foundation.

Caissons are similar to piles, except they are steel-reinforced concrete, which is poured on-site. A hole is bored in the earth (often right into the bedrock, which is referred to as a socket for the caisson to sit within), and a reinforcing cage is then lowered into the bored hole where concrete is placed. The remainder of the foundation construction is similar to that which I described for the piles. Figure 12.2 shows a simple caisson foundation.

In the case of homes built above these materials, typically the concrete slabs you see in the basement are a floating slab. The slabs carry no building loads and are designed to float up and down with the movement of the expansive soils beneath them. Some of those slabs may float as much as two to three inches between the high points in wet conditions and the low points in dry conditions.

Building a structure on top of these slabs could cause great structural damage to your home during a wet period if the new structure came into contact with the building structure above while lifting. Picture a slab lifting 3" and having the walls resting on it lifting your first floor joists that same 3".

This is a disaster waiting to happen. It is just another reason for getting professionals involved during the design stage of your project to make certain that you are not going to perform work that can damage your home or perhaps even destroy it.

Chapter 12 Codes, Permits, and Special Needs

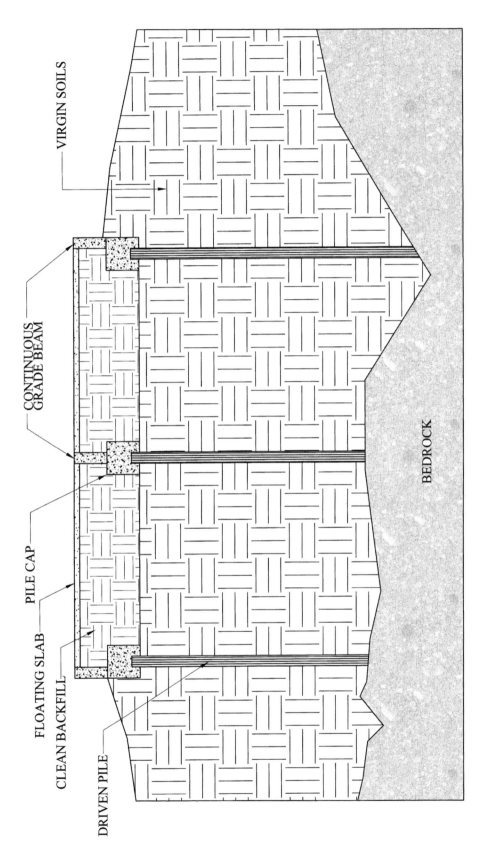

Figure 12.1 Pile foundation.

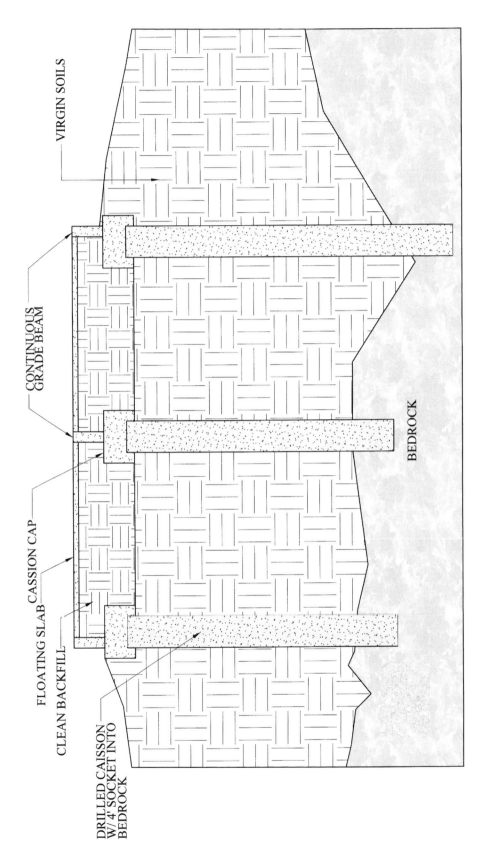

Figure 12.2 Simple caisson foundation.

Unstable Conditions Caused by the Addition of Bearing Walls

There are many other ways in which you could damage your home without realizing it. People mistakenly have the impression that bearing walls only exist if they are required to carry a load. Nothing could be farther from the truth. Any wall that is built in contact with the joists above it becomes a bearing wall, regardless of whether it needs to be or not.

Floor joists deflect under load conditions and then return to their normal condition when the load is removed. Figure 12.3 shows a floor joist in both conditions. The deflection is exaggerated so that you can see more easily what I'm describing.

Figure 12.3
Typical floor joist.

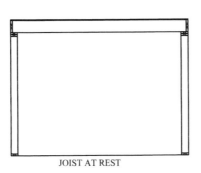

JOIST AT REST

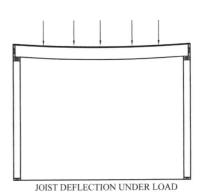

JOIST DEFLECTION UNDER LOAD

In the case of simple framing members (for example, wood floor joists), this generally will not be a problem. However, in the case of floor trusses or composite floor joists, this could be a disaster waiting to happen.

The members of a floor truss are under constant compression and tension from live and dead loads placed upon them. Figure 12.4 shows the forces in play with a simple floor truss.

If you introduce a bearing point beneath them in a location for which the truss was not designed, you could cause a collapse of the truss. Figure 12.5 shows that same floor truss with some new forces in play from a wall introduced below. Introducing a bearing wall in the location indicated in Figure 12.5 causes stresses that could very well collapse the truss. Note that the reaction of the joist (with a wall placed below) is to create a condition where the loading of the web connection is no longer symmetrical.

This creates two conditions that could cause truss failure. The first can be seen in the large-scale detail in Figure 12.5. It's the creation of a hinge point located in the bottom chord that centers in the truss plate. This causes a rotational force in the truss plate, which may cause plate failure.

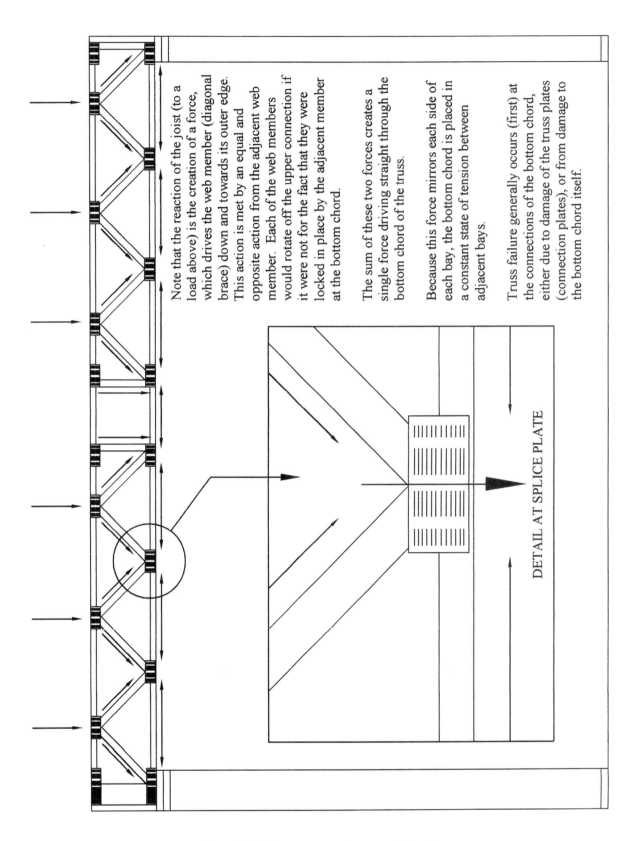

Figure 12.4 Floor truss under normal loading.

Chapter 12 Codes, Permits, and Special Needs

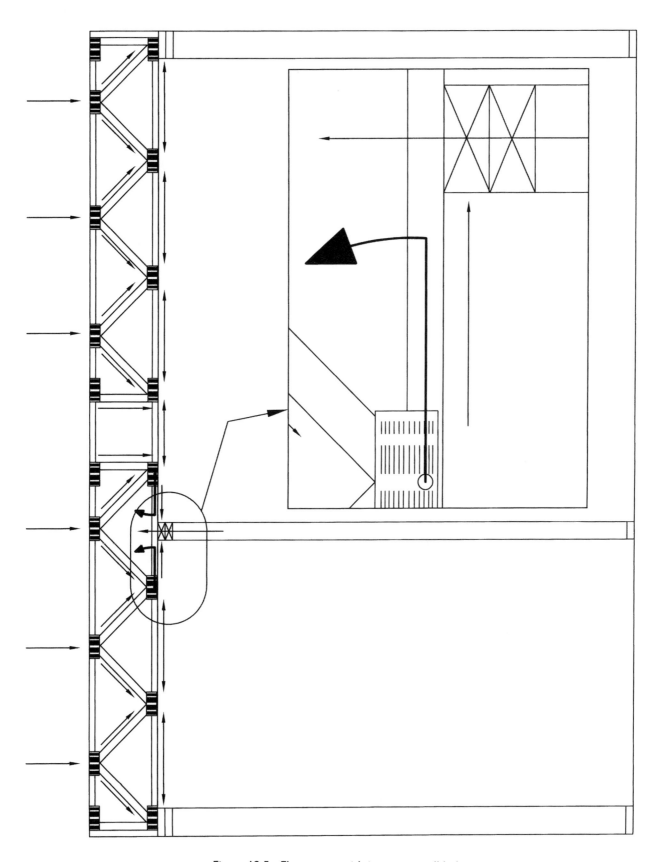

Figure 12.5 Floor truss with improper wall below.

The second condition is the loading on the bottom chord itself. A floor truss is typically constructed using 2x members installed on the flat for the top and bottom chords. The total loading is designed to carry from end to end.

If the truss we see is 20' long and installed on 2' centers, with standard design loading of 10psf dead load for truss weight and deck loading of the top chord, 10psf dead load for ceiling, mechanical, and electrical/mechanical/plumbing loads on the bottom chord, and 40psf live load, then the total load per truss would be 1,200 pounds. That would mean that 600 pounds of load was transferred to each outside bearing wall.

Introduce a bearing wall in any location along the body of that truss, and you just loaded the bottom chord with 50 percent of that load. Now we have a 2x member on the flat, and we expect it to carry a load of 600 pounds. Couple the two conditions, and you have a disaster in the making.

In the case of a floor constructed with composite joists, the manufacturers of these joists have very exacting requirements for the installation of their materials. Failure to follow these installation requirements can cause the product to collapse under load.

For example, adding a new wall in a basement, beneath the location of an existing bearing wall above, creates a condition where the load above is transferred through the composite joists to the wall below. These composite joists are not made to transfer that load and may fail by having the web of the joist collapse under the pressure. Figures 12.6 and 12.7 show the before and after conditions described previously.

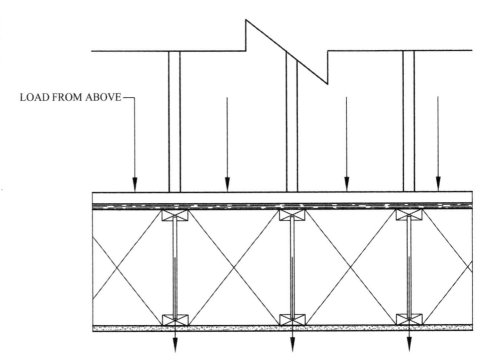

Figure 12.6 Composite joist properly loaded.

In order to install walls safely below floor trusses and composite joists, you need to understand the proper construction techniques involved and then implement them as a part of your work.

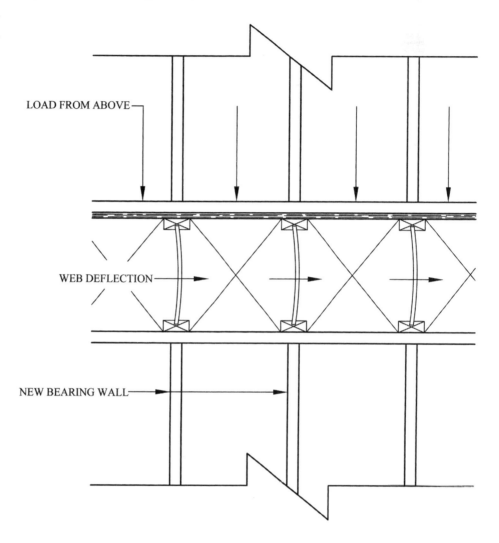

Figure 12.7
Composite joist with improper loading.

Often, you can get this information from the manufacturer (if you can identify them). If not, a structural engineer can determine the forces at work and detail the proper construction techniques required to install the new work safely. Or you can simply install all of your structure in such a manner that it never comes into contact with those members above.

My point here is that there is a lot more to this than just throwing up some walls and building a room. Sometimes, there are variables at work that you know nothing about, so it pays to spend those few extra dollars (for permits and engineering) to make certain everything will be right when you finish your work.

The examples I used here are not just thoughts of mine. They come from real-life experiences I've been involved in, and I hope they show you the importance of the message I'm sending you.

Do it right the first time, and you'll never be sorry that you did.

GLOSSARY

Glossary of Terms

A

A-B test—A test between two objects. For example, a test between two different microphones or preamps. For the test to be scientifically valid, levels should be matched, and the tests should be performed in a double-blind controlled environment. *See* double-blind.

absorption—In acoustics, the reduction in sound-pressure levels through the conversion of sound energy to heat captured within an acoustic attenuator. The opposite of reflection.

absorption coefficient—A measurement of sound energy reduction absorbed into or passed through a material. Measurement values range from 0 to 1 (which translates to 0% through 100%) and may vary with any particular material based on frequency and angle of incidence of the sound. This coefficient is typically referred to as sabin units. *See* sabin units.

acoustic impedance—A complex quotient obtained when sound pressure, averaged over a surface, is divided by the velocity of the volume through the surface.

acoustic material—Any material considered in terms of its acoustical attenuating properties. It is commonly a material designed to absorb sound.

acoustics—As used in this book, the study of the physical and psychological properties of sound as relates to isolating construction techniques and attenuation of modal and non-modal activities within a particular space.

AES—Audio Engineering Society.

airborne sound—Sound transmitted through the medium of air.

airflow resistance— The quotient of air pressure difference (steady or alternating state) across a specimen, divided by the volume velocity (steady or alternating state) of airflow through the specimen.

airflow resistivity—The quotient of the specific airflow resistance of a homogeneous material divided by its thickness.

Alcons—The measured percentage of Articulation Loss of Consonants by a listener. A % Alcons of 0 indicates perfect clarity and intelligibility with no loss of consonant understanding; 10% and beyond is increasingly poor intelligibility; 15% is typically the maximum acceptable loss; beyond 15% is unintelligible.

ambience—The acoustic characteristics of a space with regard to reverberation. A room with a lot of reverb is said to be "live"; one without reverb is said to be "dead." Note that the observation of degrees of ambiance is subjective at best, differing from person to person.

ambient noise—All-encompassing sound at a given place, usually a composite of sounds from many sources near and far.

amplitude—The magnitude of an oscillating quantity, such as sound pressure. The peak amplitude is the maximum value.

analog—An electrical signal whose frequency and level vary continuously in direct relation to the signal.

anechoic—Without reverberation.

anechoic chamber—A room designed to suppress sound reflections. Typically used for acoustical measurements.

anti-node—A resonance related to a room mode. An anti-node exists on either side of a node. Anti-nodes create an increase in amplitude, a reinforcement of a particular frequency or range of frequencies

articulation—A quantitative measure of the intelligibility of speech; the percentage of speech items correctly perceived and recorded.

articulation index—A quantitative measure of the intelligibility of speech; the percentage of speech items correctly perceived and recorded. An articulation index of 100% means that all speech can be understood and 0% means that no speech can be understood. Articulation index is calculated from the $1/3$ octave band levels between 200Hz and 6300Hz center frequencies.

artificial reverberation—Reverberation generated to simulate that of concert halls, etc. This is added to a signal to make it sound more lifelike.

Arithmetic mean sound-pressure level—The sum of sound-pressure levels, in a particular frequency band, divided by the number of measured levels from different positions, times, or both.

ASA—Acoustical Society of America.

attenuate—To alter the level of an acoustical signal through absorption, diffusion, electronic, or other means.

audible frequency range—The range of sound frequencies heard by a human ear. The audible range typically spans from 20Hz to 20,000Hz.

auditory area—The sensory area lying between the thresholds of hearing and pain.

average room absorption coefficient—Total room absorption in sabins, divided by total room surface area in consistent units of measurement.

average sound-pressure level in a room—Ten times the logarithm of the ratio of the space and time average of squared sound pressure to the squared reference sound pressure, the space average being taken over the total volume of the room, except for the regions of the room where the direct field of the source and the near field of the boundaries are of significance. Unit: decibel (dB).

A-weighting—The A-weighting curve is a wide bandpass filter centered at 2.5kHz, with -20dB attenuation at 100Hz, and -10dB attenuation at 20kHz. It exhibits heavily roll-off at the low end and a less aggressive roll-off on higher frequencies. This rating is the inverse of the 40-phon equal-loudness of the Fletcher-Munson Curve.

axial mode—Room resonances associated pairs of parallel or non-parallel walls or ceiling/floor.

B

background noise—Existing noise within a space, from sources unrelated to a particular sound, or sounds, that are the object of interest.

bandwidth—The frequency range of a system. Generally specified by establishing an upper and lower frequency range, i.e., 20–20,000Hz plus or minus 4dB.

bass—The range of audible frequencies below 250Hz.

boomy—A listening term that refers to excessive bass response.

bright—A listening term, which generally refers to upper frequency energy.

broadband noise—A wideband spectrum consisting of frequency components, no one of which is individually dominant.

C

cavity—A space between the surfaces of walls, ceilings, floors, etc., where insulation is typically installed.

characteristic impedance of the medium—The specific normal acoustic impedance at a point in a plane wave in a free field.

clipping—A type of distortion that occurs when an amplifier is driven into an overload condition. A "clipped waveform" generally contains an excess of high-frequency energy.

coherence—A listening term that refers to how well integrated the sound of a system is.

coloration—A listening term that refers to adding to a sound, something that was not in the original sound. Coloration can be a factor of room characteristics, a signal imparted by the gear being utilized in the sound reproduction, or a combination of the two.

comb filter—A distortion produced by combining an acoustical signal with a delayed replica of itself. The result is constructive and destructive interference, which results in a series of peaks and nulls introduced into the frequency response.

compression—The portion of a sound wave in which molecules are pushed together, forming a region with higher-than-normal atmospheric pressure. Acoustic compression (within a room or space) can be caused by room modes, as well as non-modal activity.

critical band—Frequency components within a narrow bandwidth that mask a given tone. Critical band varies with different frequencies, but is commonly found between 1/6 and 1/3 octaves.

critical distance—The distance from a sound source at which direct sound and reverberant sound are at the same level.

critical frequency—The frequency below the separating point where standing waves will cause significant room modes.

cutoff frequency—The frequency of a sound attenuator above which the normal incidence sound-absorption coefficient is most effective.

C-weighting—A weighting curve designed into filters for equipment measuring sound output levels. The C-curve is "flat," with -3dB corners at 31.5Hz and 8kHz, respectively. It is intended to loosely correspond to how people perceive sound at higher volume levels.

cycles per second—The frequency of a wave measured in Hertz (Hz).

D

damp—To cause a loss or dissipation of the oscillatory or vibrational energy of a signal or object. In the case of a membrane (i.e., a wall surface, the face of a panel trap, etc.), the placement of insulation touching the inner face of the membrane will create this effect.

dB—*See* decibel.

dB (A)—*See* A-weighting

dB (C)—*See* C-weighting

decay rate—As relates to airborne sound, the rate of decrease of the level of vibratory acceleration, velocity, or displacement, measured after the initial excitation has ceased.

decibel—dB, unit used to measure the amplitude of sound. A measure of sound intensity as a function of power ratio, with the difference in decibels between two sounds being given by dB=10 log10(P1/P2), where P1 and P2 are the power levels of the two sounds. The faintest audible sound, corresponding to a sound pressure of about 0.0002 dyne per sq. cm, is arbitrarily assigned a value of 0dB.

diaphragm (also diaphragmatic)—As used in this book, room surfaces or diaphragmatic sound attenuators (panel traps) that vibrate in response to sound.

diffraction—A change in the direction of sound energy in the area of a boundary discontinuity, such as the edge of a reflective or absorptive surface.

diffuse field—An environment in which sound-pressure levels are the same throughout and the flow of energy is equally probable in all directions.

digital—A converted numerical representation of an analog signal.

distortion—As used in this book, an undesired change in the waveform of a signal.

double-blind—A testing procedure, as relates to recording equipment, designed to eliminate biased results, in which the identity of the equipment is concealed from both administrators and subjects until after the study is completed.

E

echo—A reflected sound that arrives at the ear long enough after the original source to be perceived as a separate image.

Early Early Sound (EES)—Sound travels faster through a dense material than through the air. Thus, it is possible for a sound to reach a microphone through a building structure before reaching it through the air.

EFC—Energy-frequency curve.

EFTC—Energy-frequency-time curve.

equal loudness contour—For all audible frequencies, a contour representing a constant amplitude.

ETC—Energy-time curve.

extension—The extent of the range of frequencies that a device can reproduce accurately.

F

far field—That part of the sound field in which the decrease in sound pressure is inverse in relation to the distance from the source. This corresponds to a reduction of 6dB for every doubling of distance. *See* inverse-square law.

FFT—Fast Fourier Transform.

field sound transmission class, FSTC—Sound transmission class calculated in accordance with ASTM E 413.

field transmission loss, FTL—Sound transmission loss measured in accordance with ASTM E336

flame spread—The time it takes for flame to spread, measured in accordance with a variety of standards based on the use of the materials involved.

flanking transmission—Transmission of sound, from the source to a receiving location, through a structural path other than the one anticipated.

Fletcher-Munson Curve—Equal-loudness contours indicating sound-pressure levels (measured in decibels), over the range of audible frequencies, which are perceived as being of equal loudness (measured by Fletcher and Munson at Bell Labs in 1933).

flutter echo—A distinctive ringing sound caused by echoes bouncing back and forth between hard parallel surfaces following a percussive sound.

Fourier analysis—The application of the Fourier transform to a signal to determine its spectrum.

free field—Any environment in which a sound wave may travel in all directions without obstructions or reflections.

frequency—As used in this book, the number of waves (pressure peaks) that travel a distance in one second measured in Hertz (Hz).

frequency response—As used in this book, the frequency within the human hearing range, roughly 20Hz to 20kHz.

FTC—Frequency-time curve.

fundamental—The lowest frequency of a complex tone.

fusion zone—A term evolving from misinterpretations of the Haas effect. It implies that all sounds arriving (at the human ear) within an interval of time (ranging from about 20 to 50ms) are fused into a single sound. Simplistic at best, totally wrong at worst, this does not take into account that within these time limits there are several levels of clearly distinguishable characteristics, including spaciousness, timbre change, image shift, multiple sound images, and, at large delays, echoes that can color the sounds to create a separateness between them.

G

grating, diffraction—A series of minute, parallel lines used to change the direction of a sound wave.

grating, reflection phase—A diffuser of sound energy using the principle of diffraction grating.

H

Haas effect—A psychoacoustic effect, also referred to as *precedence effect*. If a listener hears two identical sounds (i.e., identical soundwaves of the same intensity) from two separate sources, then the source that is closer to the listener appears to be the only source of the sound. The level of the delayed signal may be up to 10dB higher than the direct signal before disturbing the localization effect.

hard room—A room in which the surfaces are very reflective.

harmonics—Tones at frequencies that are multiples of a fundamental tone. These are further characterized as either "even-order" or "odd-order" harmonics. Even-order harmonics (second, fourth, sixth, etc.) are even octaves above the fundamental tone. Odd-order harmonics (thirds, fifths, sevenths, etc.) are simple multiples of the fundamental, which may or may not produce pleasant overtones.

hertz—A measurement unit of frequency in cycles per second.

Helmholtz resonator—A tuned sound absorber.

I

imaging—The effect created by a stereo system that provides a three-dimensional re-creation of the sound recorded.

impulse—As regards acoustics, a short signal containing a "spike" input that sharply rises from zero and as abruptly decays back to zero.

impulse response—The response of a system to a short signal containing a "spike" input that sharply rises from zero and as abruptly decays back to zero. Impulse response measured in a room will indicate any series of reflections (from walls, etc.) of the direct response.

in phase—When two periodic waves of the same frequency synchronize, they are said to be "in phase." Signals in phase boost one another in amplitude.

initial time-delay gap—The span of time between the arrival of a direct sound and the first reflection of that sound.

interference—The combining of two or more signals in either a constructive or destructive manner.

inverse-square law—Under far/free field conditions, the effect that mandates a sound-pressure level decrease of 6dB for each (and every) doubling of the distance.

isolate—To reduce sound vibrations from passing through a structure or assembly.

J

JASA—Journal of the Acoustical Society of America.

JAES—Journal of the Audio Engineering Society.

K

KHz—1,000Hz.

L

law of the first wave front—*See* Hass effect

LEDE—*See* Live End Dead End.

Live End Dead End—An acoustical treatment plan for control rooms in which the front end is highly absorbent and the rear of the room is reflective and diffusive.

M

masking—The amount (or the process) by which the threshold of audibility for one sound is raised by the presence of another (masking) sound.

mass law—Simply, the greater mass a sound wave has to move to create vibration transfer, the greater the reduction of noise energy. Mass law dictates that for every doubling of mass, a 6dB increase in sound isolation will occur.

mean free path—Relating to sound waves in an enclosure, the average distance traveled between successive reflections, generally measured in relation to reverberation time (RT60).

millisecond (ms)—One thousandth of a second.

mode—A resonance based on a room's dimension, coinciding with the length of a particular frequency. *See* axial mode, tangential mode, and oblique mode.

modal resonance—*See* mode.

N

NAB—National Association of Broadcasters.

near field—The sound field located between the source and the far field. The near field exists under optimal conditions at distances less than four times the largest sound source dimension.

node—The opposite of an anti-node. When standing waves occur, there are positions in space relative to the wave, called *nodes*, at which there is no movement at all. The wave interferes with itself to create this instance of opposition (e.g., a wave reflecting off a wall and back into its own path). Nodes are spaced one-half wavelength apart. Nodes cause a decrease in amplitude or (if perfectly out of phase) a full cancellation of the signal.

NRC—The abbreviation for Noise Reduction Coefficient. A specification used to indicate the effectiveness of acoustic absorption materials. It is arrived at by averaging the sabin absorption coefficients of a material in the octave bands between 125Hz to 4kHz. The greater the number, the more sound absorbed.

null—A minimum pressure region in a room or space.

O

oblique mode—A room mode involving all six surfaces, i.e., the four walls, ceiling, and floor.

octave—A doubling or halving of frequency.

P

passive absorber—A sound attenuator that dissipates sound energy as heat through absorption.

PFC—Phase-frequency curve.

phase cancellation—An occurrence when two signals of the same frequency are out of phase with each other, resulting in a net reduction in the overall level of the combined signal. If two identical signals are 180 degrees out of phase, they will completely cancel one another if combined.

phon—A unit of perceived loudness, a subjective measure of the strength (not intensity) of a sound. At a frequency of 1kHz, 1 phon is defined to be equal to 1dB of sound-pressure level above the nominal threshold of hearing.

pink noise—Random noise produced with equal energy per octave.

plenum—A large (typically), absorbent-lined duct through which conditioned air is routed. Plenums are generally used as a supply (or return) connection for multiple branch ducts.

psychoacoustics—The study of the effect of the human auditory system and brain as it relates to acoustics.

pure tone—A tone without harmonics.

R

random noise—A noise signal, commonly used in measurements, which has constantly shifting amplitude, phase, and a uniform distribution of energy.

rarefaction—The portion of a sound wave in which molecules are spread apart, an expansion of the sound pressure field. The opposite of compression.

reactance—The opposition to the flow of electricity posed by capacitors and inductors.

reactive absorber—A sound absorber, such as the Helmholtz resonator, which involves the effects of mass and compliance as well as resistance.

receiving room—As relates to acoustical measurements, the space in which sound transmitted from a source room is measured.

reflection—As relates to sound, a return of residual sounds after striking a surface within a room or space. Reflection in low frequencies does not truly exist in the sense of ray tracing. Higher frequencies will display off a surface, although the strongest signal will be returned based on the theory that the angle of reflection equals the angle of incidence

reflection-phase grating—A sound diffuser using the principle of diffraction grating.

resonance—Resonance is the tendency of a mechanical or electrical system to vibrate or oscillate at a certain frequency when excited by an external source, and to keep oscillating after the source is removed.

resonant frequency—The frequency at which resonance occurs.

reverb—The remainder of sound that exists in a room after the source of the sound has stopped.

RFZ—Reflection-free zone.

room mode—The normal modes of vibration of an enclosed space. *See* mode.

RT60—Reverberation time measured as the time required for a reduction of sound pressure levels by 60dB.

S

Sabine—Wallace Sabine, the originator of the Sabine reverberation equation.

sabin unit—A unit of acoustic absorption equivalent to the absorption by one square foot of a surface that absorbs all incident sound.

sound absorption—The property possessed by materials, objects, and structures, such as rooms of absorbing sound energy, measured in sabin units.

sound absorption coefficient—The absorptive properties of a material, in a specified frequency band, measured as outlined in ASTM C423; again, typically measured in sabin units.

sound-pressure level (SPL)—Ten times the logarithm of the ratio of the time-mean-square pressure of a sound, in a stated frequency band or with a stated frequency weighting measured as a decibel (dB).

sound transmission class (STC)—A single number rating, calculated in accordance with ASTM E 413, using values of sound transmission loss. STC ratings are calculated based on the frequency range of human speech.

sound transmission loss (STL)—A measurement of the reduction in sound level when sound passes through a partition, floor, or ceiling assembly.

standing wave—A low-frequency resonance condition, within an enclosed space, in which sound waves traveling in one direction interact with those traveling in the opposite direction, resulting in a stable condition exhibiting a series of localized peaks and nulls.

structure-borne noise—Transmission of sound through structural members in a building.

T

T60—*See* RT60.

tangential mode—A room mode produced by reflections off four of the six surfaces of the room.

TDS—Time-delay spectrometry.

TEF—Time, energy, frequency.

tuning frequency—The resonant frequency of a tuned sound attenuator.

U

ultrasonic frequency—The frequency range that exists and is higher than the nominal frequency range of human hearing, measures as hertz (Hz).

V

vibration—Oscillation of a parameter that defines the motion of a mechanical system.

vibration isolation—A reduction in the ability of a structural system to transmit vibration in response to mechanical excitation, attained through the use of a resilient coupling or any other manner of decoupling of separated structural assemblies.

W

watt—A measurement unit of electrical or acoustical power.

wavelength—The distance a sound wave travels to complete one cycle. The wavelength of any frequency is found (in the simplest sense) by dividing the speed of sound by the frequency.

white noise (ANS)—Noise with a continuous frequency spectrum and with equal power per unit of bandwidth.

white noise—Random noise with equal energy per frequency.

APPENDIX

Online Tools

Folks, you were promised from the very start of this book that you would not have to do any real math in order to build your studio. As I worked my way through the book, I realized that this would be more difficult than I imagined unless you were provided with the tools you needed to accomplish this.

The available tools were written by me (the author); Brian Ravnaas, the head of engineering at Audio Alloy; and Jeff D. Szymanski, PE, the former head of engineering at Auralex Acoustics. If you want a copy of any (or all) of these tools, simply shoot a quick email to rgervais10@hotmail.com and let me know what you're looking for. I will happily send you what you need.

My personal thanks to both Brian and Jeff for their efforts in this regard. They constantly give of themselves at Internet Web sites to help people understand acoustics more clearly. Their assistance in putting together the toolkit for this book just shows again their devotion to both the field of acoustics and to you, the readers.

The available tools include the following files:

- Panel Absorber.xls, a simple calculator for a narrowband panel trap, written by the author.
- Fmsm_calc_RG.xls, a calculator to determine the Mass Air Mass frequency of various wall assemblies you might consider constructing, written by Brian Ravnaas.
- OC 703-705.xls, a spreadsheet of the Owens Corning 703/705 line of rigid insulation for use with some of the toolbox tools. Please build on this for your use, provided by the author.
- TLmodeler_mini_RG.xls, a neat little tool for analyzing Transmission Loss data in wall assemblies, written by Brian Ravnaas.
- HelpTLmodeler_mini.pdf, a help file for use with the preceding tool.

- Sabin - RT60 calculator.xls, a simple calculator for determining your future room's RT60, both empty and after treatments, and a good tool for helping you develop a treatment budget during the design process, written by the author.

- HVAC Calculator.xls, a calculator to determine sensible and latent HVAC loads in your studio, written by the author.

- Quadratic Diffusor.xls, a calculator for "Well-Type Quadratic Diffusors," ranging to prime 31. Includes a side view and inverted 3D view of your diffusor, written by the author.

- Resonance_toolbox.xls, this is a great little tool by Brian Ravnaas that calculates single-panel isolation values based on Mass law, and also includes a modulus and critical-frequency calculator for free panels, a bound panel resonance calculator, and a multi-layer panel stiffness calculator.

- Helmholtz Calculator.xls, called "Helmholtz Calculator," a simple calculator for slat-type Helmholtz traps, written by the author.

- Room Mode Calculator.xls, called "Room Mode Calculator," written by Jeff D. Szymanski, PE, this is probably the best room mode calculator I have ever worked with. If you want to calculate it, this tool gives it all to you.

INDEX

A

"A" room design (Power Station), 3–10
absorption
    ASTM C423, 199–200
    fiberglass panels, 202–203
    mid/high-bass absorber, 214
absorption coefficient, 199–201
absorption control
    fabric, 198
    fiber, 197
    foam, 198
    room treatment, 197–198
Acousti Soft, Inc., 162–163, 177
Acoustic Absorbers and Diffusers, 201
acoustic ceiling, 8
Aerosmith, 2
AES (Audio Engineering Society) standards, 121
air conditioner. *See also* cooling/heating systems
    evaporative coolers, 145–146
    exchange chambers, 146–147
    mini-split system, 144
    portable, 144–145
    split/packaged direct-expansion (DX), 140–142
    through-the-wall system, 142–143
air-handler units, 12
air space, floating concrete slab, 54
Airborne sound transmission loss test, 57
airtight construction, 46
amplitude, 23–24
antinodes, 25
apartment/condo, 15–16
ASHRAE (American Society of Heating, Refrigerating, and Air-Conditioning Engineers), 135
assembly, 43–44
ASTM C423, 199–200
ASTM E 84, 201
ASTM E 90, 41
ASTM E 413, 41

asymmetrical room design, 13
attic, 274–275
Audio Engineering Society (AES) standards, 121
audio signal ground, 118
Audio Tile ShockWave, 230–231
Auralex products, 228–231
Avatar Studios, 2
axial modes, 34–35

B

back wall interference, 29
baffle system, 154–155
"balanced load," 115
barrel roll effect, 11
basement layout, 240–241, 245–246
bass traps
    low-frequency control, 194–195
    room design, 13
bearing capacity, 319–320
bearing walls, unstable conditions caused by, 323–327
Benedictus, Edouard, 95
Berger, Russ, 230
board spacing, 4–5
Body Surface Area (BSA), 131–132
Bon Jovi, 2
Bongiovi Entertainment, 2
Bongiovi, Tony, 2
branch ducting, 154
breaker, 115
BSA (Body Surface Area), 131–132
Btu (British thermal unit) output, 131
building code and permits
    building officials, 318
    structural analysis of proposed work
        bearing capacity problems, 319–320
        unstable conditions, 323–327

C

CADD program, 242
caissons, 320, 322
calculator, mode, 37
carcinogens, 300–302
caulk
    fire, 258
    window frame construction, 98
ceiling
    adding mass to existing deck, 76
    barrel roll effect, 11
    bridging, 77
    dropped, 263
    existing ceilings, 73–78
    height, 256
    independently framed, 85–86
    resilient channel, 78–81
    semi-independent frame, 82–84
    suspended, 82
    WHR ceiling hanger, 82
ceiling clouds, 224–226
ceiling reflections, 293
channel level indicator, 180
cinder block, 288
circuit diagram, 119
CLD (constrained layer damping) system, 88–89
close micing, 26
coffin-shaped compression ring, 3–4
coincidence, 44
coincidence effect, 95
comb filtering, 8, 30–31
computer room, 13
computer-based audio analysis, 166
concrete, 73
concrete block, 72
concrete slabs
    field sound transmission loss value tabulation, 58–59
    floating, 52–56
    Impact Noise Rating (INR), 58–59
    isolated, 49–51
    Mason jack-up floor slab system, 52–53
    simple, 48–49
    test methods, 56–60
condo/apartment, 15–16

conducted RFI, 125
constrained layer damping (CLD) system, 88–89
constructive and destructive action, 25
continuation lines, 272
contractor, 20–21
control room design, 8–9, 12–14
cooling/heating system, combination. *See also* air conditioner; HVAC systems
    electric heat air conditioner, 151
    exchange chambers, 151–152
    packaged terminal air conditioner (PTAC), 151
    split systems with DX coils, 148–149
    split systems with heat pumps, 150
    through-the-wall systems, 150–151
corners, 291–292

D

dampers, 167
damping systems, 88–89
D'Antonio, Peter, 216
Dark Pine Studios, 12–15
dB (decibel), 43
dead load calculation, 256–257
dead rooms, 290
deck assembly, 242–243
deck, wood, 60–62
decoupled floor system, 269–271
design. *See* room design; studio design
Dietrich Trade Ready Design Guide, 72
diffusor, 156
    do-it-yourself room treatment, 215–221
    polycylindrical, 217–221
    Quadratic Residue, 215–217
    RPG Diffusor Systems, 216
    SpaceArray, 228–229
Digital Signal Processing (DSP), 191
dim lighting systems, 124, 126
DNSB sway brace, 87
do-it-yourself room treatment
    best use of space, 221, 224
    ceiling clouds, 224–226
    diffusors, 215–221
    fiberglass panels, 202–207
    fire test standard, 201
    foam, 202

Helmholtz traps, 210–213
low-bass panel traps, 208–210
door construction
door closure, 105
door stop, 105
double-door assembly, 107
drop seal assembly, 109
frame, 104–105
gasket, 108
hardware, 108
hinge, 105
insulation panels, 109–111
jambs, 105
latch, 105
manufactured doors, 112
Overly Door Company, 103, 112
overview, 103
seals, 108–109
sheet lead, 105
"super door," 104–105
tempered glass, 95
typical door assembly, 106
weather stripping, 105, 108–109
windows in doors, 108
double-blind listening test (Floyd), 162
Douglas Fir lumber, 277
drafting, 242
drop seal assembly, 109
dropped ceiling, 263
drywall load, 258
DSP (Digital Signal Processing), 191
dual wood-framed wall assemblies, 68–71
duct layout example, HVAC systems, 48, 245–247, 249–250, 252, 2472
Dupont Neoprene pad, 53
DX coils, split systems with, 148–149

E

early reflections, 164–165
room treatment, 197–199
and stereo imaging, 31
Early Sound Scattering (ESS), 290
EBU (European Broadcasting Union), 295
egg crates, 303–309
electric heat air conditioner, 151

electrical considerations
amperage load, 117
audio signal ground, 118
"balanced load," 115
breaker, 115
circuit diagram, 119
diagnosing and troubleshooting problems, 126–129
electrical noise, 118–121
electrical panel, 115–116
ground loops, 118–121
grounding, 118–121
isolated ground receptacles, 122
lighting, 124
line voltage, 114–117
low-voltage wiring, 117
radio frequency interference (RFI), 125–126
star grounding system, 122–123
sub-breaker configuration, 115
elevations and sections through rooms, 254–266
Energy Time Curve (ETC), 167–168
Energy-Time graph, 164
equalization, 190–192
equipment options, 1
ESS (Early Sound Scattering), 290
ETC (Energy Time Curve), 167–168
European Broadcasting Union (EBU), 295
evaporative coolers, 145–146
EveAnna Manley, 1
Everest, F. Alton (*The Master Handbook of Acoustics*), 300
Everett, Gary, 110
exchange chambers, 146–147, 151–152
expansive soils, 319
exterior/garage wall assemblies, 271–273

F

fabric
absorption control, 198
fire retardant, 207
Guilford of Maine Fabric, 207
room design and, 9
Faraday Cage, 129
Faraday, Michael, 129
Faraday Shield, 129

fiber, 197
fiberglass panels
    absorption chart, 202–203
    backing, 204–205
    cloth covering, 207
    do-it-yourself room treatment, 202–207
    face fabric, 204–205
    fire-safing, 203
    frame with insulation, 204–205
    myths and legends, 300–302
    Owens Corning, 202
    standoffs and hangers, 206
    wood frame construction, 203–204
fire caulk, 258
fire retardent fabric, 207
fire test standard, 201–202
fire-safing, fiberglass panels, 203
Flame Stop Inc., 207
flanking path, 46–47, 257
float glass, 94
floating concrete slab, 52–56
floating frames, 258
floating walls, 276–284
floating wood deck, 60–62
floor construction
    concrete slabs
        field sound transmission loss value
            tabulation, 58–59
        floating, 52–56
        Impact Noise Rating (INR), 58–59
        isolated, 49–51
        Mason jack-up floor slab system, 52–53
        simple, 48–49
        test methods, 56–60
    decoupled floor system, 269–271
    wood deck, 60–62
floor joist, 257, 323
floor truss, 323–326
fluorescent lighting, 124
flutter echo, 29–30
foam
    absorption control, 198
    room treatments, 202
footing slab, 258
foundation/footing slab, 258
Fourier synthesis, 168

Fourier Transform, 168
frequency
    basic description of, 23
    defined, 23
    distance of travel, 24
    frequency chart, 38–40
    low-frequency control, 194–195
    resonant, 44
frequency response curve, 168
fresh air, 135–136

G

garage/exterior wall assemblies, 271–273
gasket, 108
gating, 169–170, 173
Gauss, C.F., 216
Gaussian curve, 173
Gaynor, Gloria, 2
General Motors, 108
GIK Acoustics products, 233–235
glass
    float, 94
    heat strengthened, 94–95
    laminate, 95–96
    molton, 94
    plexiglass, 96
    tempered, 94–95
    thickness, 100–101
glazing tape, 98
grass, 169
Green Glue, 88–89
ground loops, 118–121
ground receptacles, 122
grounding, 118–121
grout versus sand, 289
Guilford of Maine Fabric, 207
gypsum board, 44
gypsum concrete application, 73

H

"H" room design, 10–12
Hanning curve, 173
harmonic analysis, 168
harmonics, 44
Hass effect, 295

hat channel, 18
haunch, 49
headphone, 27
heat pump, 150
heat-strengthened glass, 94–95
heating. *See* cooling/heating systems; HVAC systems
heatload calculation, 133–134
Helmholtz traps, 210–213
Hendrix, Jimi, 2
hinge, 105
Hit Productions, Studio H, 10–12
human factor, 313–314
human hearing range, 171
humidity, 136–139
HVAC systems. *See also* air conditioner; cooling/heating systems
    duct layout example, 245–250, 252
    humidity and, 136–139
    Nailor Industries Series RBD performance data, 156–157
    noise criteria levels, 158–159
    overall functionality, 139–140
    overview, 130
    planning and design considerations, 251–254
    room design criteria
        Btu output, 131
        fresh air, 135–136
        heatload breakdown per activity, 135
        heatload calculation, 133–134
        latent loads, 135
        the people factor, 131–132
        room calculations, 132–134
        sensible loads, 134
    separate systems, 153–154
    solar power, 133
    system design, 154–157
    Ventilation Load Index (VLI), 135–136
    volume *versus* velocity, 140
hybrid devices, 197
Hybrid signal type, 181–182
hygrometer, 137–138

I

Impact Noise Rating (INR), 58–59
impact test, 56–57
imposed loads, 278
impulse response, 167
independently framed ceiling construction, 85–86
INR (Impact Noise Rating), 58–59
inspection, 20–21
insulation
    door construction, 109–111
    fiberglass panel with, 204–205
    sound-batt, 15
inter-stimulus delays (ISDs), 164
iso-booth room design, 8
isolated concrete slabs, 49–51
isolated ground receptacles, 122
isolation. *See* sound isolation
isolation transformer, 121

J

jamb, 105
Joemeek, 1

L

laminate glass, 95–96
lamp debuzzing coils (LDCs), 126
large rooms, 33
latch, 105
latent loads, 135
LDCs (lamp debuzzing coils), 126
LED (Liquid Emitting Diode) lighting, 124
LEDE (Live End/Dead End) room, 290
Legendre, A.M., 216
lehr, 94
LF (low-frequency) soundwave, 36
LF response measurement, 184–185
lighting
    dimmer systems, 124, 126
    fluorescent, 124
    lamp debuzzing coils (LDCs), 126
    Liquid Emitting Diode (LED), 124
    Lutron Electronics manufactured, 124
    mood, 124
    noise, 126

line voltage, 114–117
Liquid Emitting Diode (LED) lighting, 124
Live End/Dead End (LEDE) room, 290
Localization Dominance phase, 165
loopback test, 178
lounge, 13
low-bass panel traps, 208–210
low-frequency control, 194–195
low-frequency (LF) soundwave, 36
low-voltage wiring, 117
Lutron Electronics, Inc., 124

M

MAM (Mass Air Mass) systems, 64
Manley in Chino, 1
manufactured doors, 112
manufactured room treatments
    Audio Tile ShockWave, 230–231
    Auralex products, 228–231
    GIK Acoustics products, 233–235
    MiniTraps, 231–232
    overview, 226
    Ready Acoustics products, 236–237
    RealTraps products, 231–233
    SpaceArray diffusor, 228–229
    SpaceCoupler, 229–230
    testing facilities, 227
manufactured window units, 103
Mason Industries
    isolation products, 87
    jack-up floor slab system, 52–53
    ND Isolator, 91, 276, 280
    WHR ceiling hanger, 82
    wood-framed floating floors, 61
masonry, 72, 288–289
mass, 19, 43–46
Mass Air Mass (MAM) systems, 64
Mass Law, 43–44
mass loaded vinyl (MLV), 100
Mass Spring Mass (MSM) system, 53
Master Handbook of Acoustics, The, (Everest), 300
mattress, 311–312
M-Audio, 1
medium-sized rooms, 33

member, 18–19
microphone
    close, 26
    placement, 3–4
mid/high-bass absorber, 214
mini-split air-conditioner system, 144
mini-split heat pump, 150
MiniTraps, 231–232
mirror trick, 164
misinformation. *See* myths and legends
MLS test signal, 180
MLV (mass loaded vinyl), 100
modal waves, 25–28
modes. *See* room modes
mold, 138–139
molton glass, 94
mood lighting, 124
Mosteller formula, 131–132
MSM (Mass Spring Mass) system, 53
myths and legends
    cautions, 317
    close is good enough, 314–317
    egg crate use, 303–309
    fiberglass panel, 300–302
    mattress, 311–312
    packing foam, 311
    soundproofing, 312

N

National Council for Acoustical Consultants (NCAC), 230
ND Isolator (Mason Industries), 91, 276, 280
Neoprene Partition Supports (NPS), 87
Newton's Cradle, 27
NFPA 701 test standard, 201
nodes, 25
noise criteria levels, 158–159
non-modal waves, 28–29
NPS (Neoprene Partition Supports), 87

O

oblique modes, 35–36
open combination studio/control room, 244

Osbourne, Ozzy, 2
Overly Door Company, 103, 112
Owens Corning fiberglass panels, 202

P

packaged terminal air conditioner (PTAC), 151
packing foam, 311
panel resonance, 43
panel traps, 208–210
parameter EQ, 190–192
partition sound transmission losses, 41
percussion room, 5
Phase II ISDs, 165
piles, 320–321
plexiglass, 96
Plumb, Doug, 177
point loads, 276
polycylindrical diffusor, 217–221
portable air conditioner, 144–145
power company service, 114
Power Station Studios, 2–10
Precedence Effect, 164
PreSonus, 1
pressure devices, 196
product deficiency, 314–315
project superintendent, 19
psychoacoustics, 176
psychological response, 175
psychrometer, 137–138
PTAC (package terminal air conditioner), 151
PVC pipe, 289

Q

Quadratic Residue diffusor, 215–217
quality control, 19–21, 314–315

R

radio frequency interference (RFI), 125
Ramones, The, 2
ranch basement layout, 240–241, 245–246
ratio, room, 36–37
Ravnaas, Brian, 109
RBDG (Russ Berger Design Group), 230
RC-2, 18

RCP (reflected ceiling plan), 251, 253
Ready Acoustics products, 236–237
RealTraps products, 231–233
reflected ceiling plan (RCP), 251, 253
Reflection Free Zone (RFZ), 291, 296–297
Reflection Phase Grating Diffusor, 216
reflective problems
    back wall interference, 29
    comb filtering, 30–31
    early reflections and stereo imaging, 31
    flutter echo, 29–30
    speaker boundary interference response (SBIR), 29
relative humidity, 136–137
resilient channel
    ceiling construction, 78–81
    wall construction, 66–67
resonant frequency, 44
resonant sounds, 166
Rettinger rooms, 290
reverse cycle chiller, 150
RFI (radio frequency interference), 125
RFZ (Reflection Free Zone), 291, 296–297
rhythm room, 5
RISC assemblies, 66
roof structure, 284–288
room design. *See also* studio design
    acoustic ceiling, 8
    air-handler units, 12
    asymmetrical, 13
    bass traps, 13
    board spacing, 4–5
    coffin-shaped compression ring, 3–4
    computer room, 13
    control room, 8–10, 12–14
    Dark Pine Studios, 12–15
    fabric use, 9
    functionality concerns, 3
    Hit Productions (Studio H), 10–12
    iso-booth, 8
    lounge, 13
    mic placement, 3–4
    percussion/rhythm room, 5
    Power Station, (the "A" room), 3–10
    room geometry, 5
    room ratio, 36–37

string room, 6–7, 11
symmetrical, 8–9
tracking room, 12–14
wood finish, 11
Room EQ Wizard program, 163
room modes
  amplitude and, 24
  axial modes, 34–35
  basic description of, 23
  equation, 24
  frequency chart, 38–40
  frequency's distance of travel, 24
  modal waves, 25–28
  mode analysis, 32–33
  mode calculators, 37
  non-modal waves, 28–29
  oblique, 35–36
  reflective problems and, 29–31
  room size, 33–34
  tangential modes, 35
room testing
  channel level indicator, 180
  data gathering, 179–182
  double-blind listening test (Floyd), 162
  early reflections, 164–165, 188–189
  frequency response curve, 168
  gating, 169–170, 173
  harmonic analysis, 168
  Hybrid signal type, 181–182
  impulse response, 167
  loopback test, 178
  nature of scientific measurement and experiment, 176–178
  operation of analyzers
    isolation between spaces, 189–190
    LF response measurement, 184–185
    precise and accurate measurement, 183–184
    speaker power response, 186
    SPL data gathering, 186–187
  parametric EQ, 190–192
  psychological response, 175
  resonant sounds, 166
  room anomalies, 163–165
  signal-to-noise ratio, 181–183
  software, 162–163
  sound card calibration, 179
  Sweeps signal type, 181–182
  waterfall response, 170–175
room treatment
  absorption coefficient, 199–201
  dead rooms, 290
  do-it-yourself treatment
    best use of space, 221, 224
    ceiling clouds, 224–226
    diffusors, 215–221
    fiberglass panels, 202–207
    fire test standard, 201
    foam, 202
    Helmholtz traps, 210–213
    low-bass panel traps, 208–210
    mid/high-bass absorber, 214
  early reflection control, 197–199
  Early Sound Scattering (ESS), 290
  hybrid devices, 197
  Live End/Dead End (LEDE) room, 290
  low-frequency control, 194–195
  manufactured
    Audio Tile ShockWave, 230–231
    Auralex products, 228–231
    GIK Acoustics products, 233–235
    MiniTraps, 231–232
    overview, 226
    Ready Acoustics products, 236–237
    RealTraps products, 231–233
    SpaceArray diffusor, 228–229
    SpaceCoupler, 229–230
    testing facilities, 227
  pressure devices, 196
  Reflection Free Zone (RFZ), 291, 296–297
  Rettinger rooms, 290
  room design, 10
  velocity devices, 196
RPG Diffusor Systems, 216
RPlusD software, 162–163, 177–178
RT60 (T60), 30
rubber gasket, 108
Russ Berger Design Group (RBDG), 230

S

sabin, 200
sand *versus* grout, 289

sand-filled wooden deck, 62
Sayers, John L., 300
SBIR (speaker boundary interference response), 29
Schroeder, 215
Scorpions, The, 2
seals, 108–109
Sebatron, 1
semi-independent frame ceiling construction, 82–84
sensible loads, 134
setting blocks, 98
sheet lead, door construction, 105
signal processing, 167
signal-to-noise ratio, 181–183
Sika Corporation, 51
SikaDur Hi Mod, 51
single rooms, 242
size, room, 33–34
slab, concrete
    field sound transmission loss value tabulation, 58–59
    floating, 52–56
    Impact Noise Rating (INR), 58–59
    isolated, 49–51
    Mason jack-up floor slab system, 52–53
    simple, 48–49
    test methods, 56–60
sling psychrometer, 137
slot resonator, 210–213
small rooms, 33–34
smoke index, 202
software, room testing, 162–163
solar power, 133
sound
    amplitude, 23
    frequency, 23
    wavelength, 23–24
sound card calibration, 179
sound energy, 25
sound equipment interconnection, 121
sound isolation
    acoustic engineer to monitor, 42
    airtight construction, 46
    coincidence, 44
    condo/apartment living, 15–16
    eliminating transmissions through building structure, 46–47
    flanking paths, 46–47
    isolation transformer, 121
    for lower frequencies, 43
    mass conditions, 43–46
    Mass Law, 43–44
    panel resonance, 43
    partition supports, 87
    resonant frequency, 44
    sound level meters, 42
    sound-batt insulation, 15
    STC ratings, 41
    what to avoid, 47
sound level meters, 42
Sound Transmission Class (STC) rating, 41, 62–63
soundproof. *See* sound isolation
soundproofing myth, 312
soundwave, 23, 36
SpaceArray, 228–229
SpaceCoupler, 229–230
speaker
    location layout, 293–294
    power response measurement, 186
speaker boundary interference response (SBIR), 29
speed of sound, 23–24
split system
    with DX coils, 148–149
    with heat pumps, 150
split/packaged direct-expansion (DX) air conditioner, 140–142
star grounding system, 122–123
STC (Sound Transmission Class) rating, 41, 62–63
steel framed wall construction, 71–72
stereo imaging, 31
storage room, 245
string room, 6–7, 11
structural analysis of proposed work
    bearing capacity problems, 319–320
    unstable conditions, 326–327
studio design. *See also* room design
    CADD program, 242
    ceiling height, 256
    dead load calculation, 256–257
    deck assembly, 242–243
    drafting, 242

elevations and sections through rooms, 254–266
floating frames, 258
floor joist, 257
foundation/footing slab, 258
HVAC systems, 251–254
multiple rooms, 242
open combination studio/control room, 244
ranch basement layout, 240–241, 245–246, 248–250
same room, 244
separate room, 244
single rooms, 242
storage room, 245
symmetry, 245
truss, 260
wall construction, 257
Studio H, 10–12
Summing Localization phase, 165
"super door," 104–105
suspended ceiling construction, 82
sway braces, 87–88
Sweeps signal type, 181–182
symmetrical room design, 8–9
symmetry, 245

T

T60 (RT60), 30
Talking Heads, 2
tangential modal wave, 25–26
tangential modes, 35
Tascam, 1
technological advances, 1
tempered glass, 94–95
testing. *See also* room testing
    Airborne sound transmission loss test, 57
    concrete slabs, 56–60
    impact, 56–57
T&G (Tongue and Groove), 73
tin roof, 12
TJI (Truss Joist), 73
Toft, Malcolm (Toft Malcolm Designs), 1
tone, 23
Tongue and Groove (T&G), 73
Toole, Floyd (double-blind listening test), 162
tracking room, 12–14

treatments. *See* room treatment
Trident Boards, 1
Triplex, 95
Tripp Lite, 128–129
troubleshooting, 126–129
"trough" areas, 25
truss, 260
truss failure, 323
Truss Joist (TJI), 73

U

UCONN Web site, 137
underpinning, 50
unstable conditions, 323–327
USG Gypsum Construction Handbook, 72

V

velocity
    defined, 23–24
    devices, room treatment, 196
    volume *versus,* 140
VLI (Ventilation Load Index), 135–136
voltage
    line, 114–117
    low, 117
    power company service, 114
volume, 140

W

wall braces, 87–88
wall construction
    existing walls, 64–65
    floating walls, 276–284
    masonry, 72
    planning and design considerations, 257
    STC ratings, 62–63
    steel framing, 71–72
    two-leaf system, 64
    wood
        dual-frame assemblies, 68–71
        resilient channel/RISC assemblies, 66–67
        single wall construction, 66
waterfall response, 170–175

wavelength
- basic description of, 23–24
- equation, 23
- frequency chart, 38–40

WCL sway brace, 87
weather stripping, 105, 108–109
Web site resources, 300
WHR ceiling hanger, 82
WIC sway brace, 87
window frame construction
- black felt use, 98
- caulk, 98
- double-glazed window assembly, 96–99
- glass thickness, 100–101
- glazing tape, 98
- manufactured units, 103
- separated wood frame assembly, 96
- setting blocks, 98

window, in door, 108
Winer, Ethan, 163, 291
wire rope, 225
wiring. *See* electrical considerations
wood deck, 60–62
wood finish, room design, 11
wood wall construction
- dual-frame assemblies, 68–71
- resilient channel/RISC assemblies, 66–67
- single wall, 66

Z

Zero International, 108